DESIGN
D&T
MAKE IT!

product design KS4

D1413033

Brian Russell ■ **Tristram Shepard**

Nelson Thornes
a Wolters Kluwer business

Published in 2006 by:
Nelson Thornes Ltd
Delta Place
27 Bath Road
CHELTENHAM
GL53 7TH
United Kingdom

06 07 08 09 10 / 10 9 8 7 6 5 4 3 2 1
A catalogue record for this book is available from the British Library
ISBN 0 7487 9791 2

Edited by Judi Hunter
Illustrations by Tristram Arris, Stuart Douglas, Hardlines, Andrew Loft, Tristram Shepard
Picture research by John Bailey, Tristram Shepard, Turchini Design Ltd
Designed and typeset by Turchini Design Ltd
Printed and bound in Croatia by Zrinski

Acknowledgements

Thanks are due to Melanie Fasciato and Alex McArthur, and to Tony Wheeler, Alex French, Sally White, Liz Cook, Ross McGill and Will Potts for their contributions, comments and suggestions. I am also grateful to Stuart Douglas from Ripley St. Thomas CofE High School for supplying examples of project work from Kezi Lailey, Viki Peills and Huw Roberts, and for the case study on the work of douglasarkwright. Nick Bradbury kindly provided the CAD images produced by his students at Edensnor Technology College on page 91. The photographs of Susannah Smith's soft models on page 60 were taken at Saltash.net Community School.

The sections on soft modelling, the use of handling collections and student collaboration have been developed from the work of Richard Kimbell, Tony Wheeler and the TERU team at Goldsmiths College as part of the e-scape project.

The text on pages 22–23 was based on 'Blobitechture' by John K Waters, published by Rockport, 2003, and a presentation given by Bruce Stirling (www.boingboing.net/images/blobjects.htm).

The 'What's the score?' text on page 15 was based on a system devised by Edwin Datschefski in 'The Total Beauty of Sustainable Products', published by Rotovision (2001).

... was based on a text ... an article by Sheryl ... ber 2005/News

With thanks to the following for permission to reproduce photographs and other copyright material in this book:

Adidas: p98 (lower middle right); Adobe: p10 (bottom right); AEG: p18 (top left); Alex French: p60 (bottom right); Apple: pp22 (top right), 98 (bottom middle), 130–131; Art Directors and TRIP: pp31 (middle right), 89 (bottom left), 122 (middle right), 140 (middle left); BBC: p111 (bottom left); Bridgeman Art Library Private Collection: p16 (bottom right); BSI: p124 (bottom left); Casio: p129 (top left); Child's Play: p106 (top right); Corel Corporation: pp40 (top middle and right), 42 (bottom right); Dairy Crest: p8 (middle left); David Winter: p114 (top right); Elise Co and Nikita Pashenkov of Aeolab LLC: p129 (bottom); exoteric.roach.org: p140 (bottom right); Eye-TheatreTV: p129 (middle right); FAIRTRADE: p78 (bottom right); Fantasy Flight: p111 (bottom right); Flexon: p141 (bottom right); GABON fabric. Design by N Du Pasquier for MEMPHIS MILANO collection in 1982: p21 (bottom right); Getty Images/Telegraph Colour Library: p23 (top); Hemera Technologies Inc: pp7 (top left), 8, 9 (bottom right), 12 (top left), 13 (middle left and middle right), 15 (bottom left), 17 (bottom right), 19 (top right), 21 (top right), 22 (top left and bottom left), 24 (top right), 25, 26, 27, 28, 29, 30, 31 (middle left and top right), 32, 33, 34 (top and bottom right), 35 (bottom right), 36 (top), 39 (top right), 40 (bottom right), 41 (bottom right), 43 (top, middle bottom right and bottom right), 45 (lower middle), 46, 47 (top left and bottom left), 48 (bottom left), 49, 51 (bottom left and right), 52, 53, 54, 55, 56, 57 (top), 62 (bottom right), 63 (bottom right), 64, 65, 66, 70 (bottom right), 71 (middle left), 75 (bottom left), 82 (top right), 87 (bottom middle), 89 (bottom middle), 91 (bottom), 92 (middle, bottom left and middle right), 93 (top left, top right), 101 (top right), 102 (top right), 103 (top left and bottom right), 105 (bottom left), 108 (bottom left), 109 (bottom), 110 (top right), 118 (bottom left), 120 (top right), 122 (top, middle, lower middle and bottom right), 123 (bottom), 124 (top right), 132 (top right and middle), 133 (bottom left and middle), 138 (middle centre), 144 (lower middle right), 155 (bottom left and top middle), 158 (top left); I F Machines: p140 (top right); IKEA: p98 (upper middle right); Innocent: p98 (top right); Jet Meyer: p69; London's Transport Museum © Transport for London: p18 (bottom right); Lunar Design: p141 (lower middle) Mark Bullimore: p86 (bottom middle); Martyn Chillmaid: pp35 (top and bottom left, top right), 36, 104 (top right); Mascioni: pp41 (top and middle right), 142 (bottom middle and right) ; Matthew Hull: p115 (upper middle); National Museums of Scotland: pp71 (top right), 73 (middle left and middle right); Nova Development Corporation: p24 (middle right and bottom); NTT DoCoMo Inc: p129 (middle left); Oxfam: p96; Paul Moss Photographer, Artist: p113 (middle left); Perrier name and image is reproduced with the kind permission of Société des Produits Nestlé S.A: p98; Quark: p106 (bottom left); Remarkable (Pencils) Ltd: p14 (middle) Royal Mail: p103 (bottom); Science Photo Library/Bruce Iverson: p44 (middle right), Science Photo Library/Volker steger p79 (middle left), Science Photo Library/Peter Ryan p83 (bottom right), Science Photo Library/Dr Jeremy Burgess p140 (middle left), Science Photo Library/Sam Ogden p141 (top right), Science Photo Library/Maximilian stock p142 (bottom middle); Science & Society 10316906: p20 (bottom left), and 10451731: p20 (bottom left); SIGG SWITZERLAND AG: p95; SmilePlas: p70 (top and middle right); Splash Display Ltd: p105 (middle bottom); Techsoft: p11 (bottom); Tony Wheeler: p60; topfoto.co.uk: pp18 (bottom left), 19 (bottom left), 20 (bottom right); Tristram Shepard: pp15 (middle right and upper middle), 19 (middle left), 20 (top right), 21 (top left), 22 (bottom middle left, bottom middle right and bottom right), 24 (bottom and middle left, upper and lower middle), 34 (bottom left), 57 (right lower middle), 71 (bottom left), 72 (bottom left), 87 (bottom left), 104 (bottom right and bottom middle), 128 (bottom middle), 138 (top left and centre middle left), 144 (top, upper middle and bottom left), 156, 157, 158 (middle), 159; USDA: p112 (middle right); V&A Images/Victoria and Albert Museum: p16 (top left); Vintage Calculators/Nigel Tout: p50 (bottom left); Voltaic Systems: p129 (top middle); Warner Bros: p112 (bottom left); Wild Planet: p129 (top right); www.sinclairc5.com: p50 (middle bottom); Z corporation: p11 (top right).

All other photographs are from iStockphoto.com.

Every effort has been made to contact copyright holders. The publishers apologise to anyone whose rights have been inadvertently overlooked, and will be happy to rectify any errors or omissions.

Contents

Introduction

Welcome to *Design & Make It! Product Design*. This book has been written to support you as you work through your GCSE course in Design and Technology. It will help guide you through the important stages of your coursework, and assist you in preparing for the final examination paper.

How to use this book

There are two main ways this book might be used.

1 Follow the four design and make projects in sequence over the whole course, including a selection of the suggested activities. You do not necessarily need to take all of them as far as the production of a finished working product: discuss this with your teacher.
2 Undertake alternative projects to one or more of those provided and refer to those pages which cover the specific areas of knowledge and understanding defined in the examination specification and the KS4 National Curriculum.

Contents

Introduction
The first section provides a general introduction to product design and the most important aspects of your course.

Materials and components
This reference section provides an overview of the range of materials and components you need to know about and be able to use.

If you are following a food or ceramics material route, check with your teacher exactly what you need to cover.

Project guide
The project guide summarises the design skills you will need for extended project work. Refer back to these pages throughout the course to help you develop your work. Use the

'In my design folder' checklists to ensure you are providing the evidence the moderator will be looking for.

The projects
Four coursework projects are provided. These each contain a mixture of product analysis and development pages and knowledge and understanding pages (e.g. design for manufacture). In each of the projects, the development of a possible solution has been used as an ongoing example. You could base your own work closely on this solution, but if you want to achieve higher marks you will need to try to come up with ideas of your own.

Project suggestions
Finally, two outline project suggestions are provided. You might wish to tackle these for your final coursework project.

Making it

Whatever your project, remember that the final realisation is particularly important. It is not enough just to hand in your design folder. You must have separate 3D products that you have made. The quality of your final realisation must be as high as possible, as it counts for a high proportion of the marks.

During your course you will need to develop technical skills in using a range of materials, processes and tools. This is something you can't do just by reading a book! The best way is to watch carefully as different techniques and procedures are demonstrated to you, and practise them as often as possible.

Beyond GCSE

There are good opportunities for skilled people to work in industrial production. Another alternative is to train as a product designer. Such people need to be flexible, good communicators, willing to work in teams and be computer literate.

There are a wide range of further AS, A2 and degree courses and training opportunities in Product Design available at various levels that you might like to find out more about.

IN YOUR PROJECT

The 'In Your Project' paragraphs will help you to think about how you could apply the content of the page to your current work.

KEY POINTS

Use the 'Key Points' paragraphs to revise from when preparing for the final examination paper. Two specimen papers are included on pages 108 and 154.

What is Product Design?

Design and Technology is about improving people's lives by designing and making the things they need and want. From furniture to fast food, cars to cardigans and mobile phones to mugs and plates, products are designed and made to make life easier and more enjoyable, and to make a task or activity more efficient.

Design and technology

Technology helps extend our natural capabilities. For example, it enables us to:

- travel further and faster
- send and receive messages across the world in an instant
- keep warm in winter and cool in summer.

What do designers do?

Designers help make new and existing technologies easy and more pleasant for people to use – they make them look and feel fun and fashionable, logical and safe to use.

Designing products involves working with a wide range of materials – such as wood, metal, plastic, ceramics, glass, paper, card, fibres, fabrics and food – to choose and use the ones that are most appropriate for the job they need to do.

Good Product Design involves creating something that works well and is satisfying to use. But to be successful it also has to be commercially viable, i.e. be sold at a price that will be acceptable to the consumer but also make a profit for the manufacturer and investor.

In your Product Design course

During your course you will be expected to work in similar ways to professional designers. As you develop possible solutions to complex problems you will, in particular, need to learn about:

- design methodology
- materials and components
- industrial manufacturing
- consumerism
- human factors
- packaging and labelling
- sustainability
- the use of ICT, including CAD-CAM
- the history of design.

These themes are all introduced in these introductory pages. It's very important that you demonstrate your awareness and understanding of them within your coursework.

Professional designers

- agree a brief with a client

- keep a notebook or log of all work done with dates, so that time spent can be justified at the end of the project

- understand the market for the product and keep users' needs at the forefront

- check existing ideas. Many designers re-style existing products to meet new markets because of changes in fashion, age, environment, materials, new technologies, etc.

- consider social, environmental and moral implications

- ensure legal and safety requirements are met

- set limits to the project to guide its development (design specification)

- suggest materials and production techniques that are appropriate to the number of products to be made

- produce working drawings and instructions for manufacturers to follow (manufacturing specification).

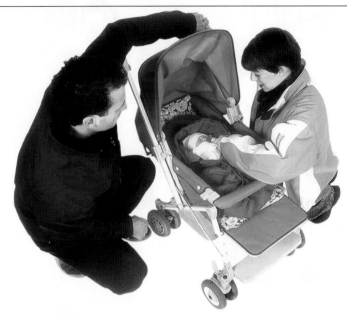

Good design involves a successful match between the available materials, tools and processes and the differing physical and emotional needs of the users.

The design process

The sequence of moving from first thoughts to having a new product on sale in the market is called the **design process**. This does *not* mean following a pre-determined series of steps. Early on, designers are likely to be more concerned about new ideas and research and, as the completion day nears, more about manufacture and presentation. In between, lots of things will be thought about and explored, discussed and developed, approved and rejected.

The exact sequence of methods used, and how long each takes, is different for each new product: it very much depends on the particular interests and strengths of the designers and the requirements and expectations of the client.

The various design methods you will need to use in your coursework are described in the Project guide section on pages 48–69.

Design methodology

Designers usually start with initial, hazy ideas and impressions about what a new product might be like. They then use a range of methods to:

■ find out more about what people need and want, and about the materials and production technologies that are available

■ come up with new imaginative and creative possibilities that might provide a better solution and appeal to potential customers
■ make drawings, sketches, models and prototypes to test out their designs until they work as required
■ present their designs in a clear, concise and convincing way to their clients and to potential manufacturers and investors.

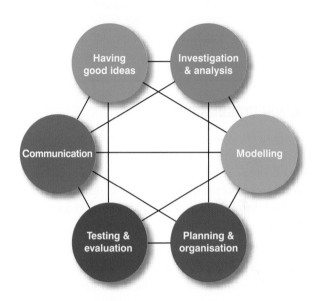

The process of design involves the frequent interaction between different skills, techniques and methods, appropriately applied to the specific problem.

Materials and components

Most materials and components behave in certain ways – they do one thing well, but another badly. Materials and components have unique **physical properties** (how they perform) and **working characteristics** (how they can be shaped and formed). Designers have to work within the constraints of the materials, components and production methods they have available.

However, new **composite** and **smart** materials provide the designer with a much wider choice, and even the possibility of creating materials and components that have unique physical properties and working characteristics to meet the requirements of a particular purpose.

There is more on properties and characteristics on pages 24–29, and on composite and smart materials on pages 140–141.

What is Product Design? [continued]

Retailers:

- need to make a profit on the products sold
- consider the market for the product to give consumers what they want, when they want it, at an acceptable price
- take account of consumers' legal rights and take consumer complaints seriously
- continually review new products
- put in place a system to replace stock levels.

Industrial manufacturing

Designers need to work within the constraints of the industrial production methods that are available to them, and the requirements of the **consumer**, **client**, **manufacturer** and **retailer**. Teamwork and good communication between these people and organisations are essential.

In industry, products are made in quantity – either in small batches or in large volumes. The quantity and variety of a product to be manufactured has a significant impact on how it can be made. Some things only become viable if made quickly and simply and on a large scale using automated manufacturing processes. If smaller quantities of a particular product are required, they may need to be made in different ways.

 There is more on industrial manufacturing on pages 63, 120–123 and 142–145.

Consumerism

How good are the products that have already been designed and made? How well do they do the jobs they say they do? **Evaluating existing products** is important for consumers and designers. This involves more than just describing a product and what it does, but also making comments on how well it works, how well it has been made and how satisfying it is to use – i.e. a mixture of facts and opinions. **Comparing** and **contrasting** (i.e. looking for things that are similar and different) two or more similar products can be very informative.

Designers and other consumer agencies constantly evaluate existing products to identify successful solutions and possible improvements.

 Throughout your course you will need to be studying existing products (see pages 72, 88,114 and 132–133).

Clients:

- identify a need or opportunity and tell designers what they want a product to do and who it is for (the brief)
- consider the possible market for the idea
- organise people, time and resources and raise finance for the project.

Consumers/users expect the product to:

- do the job it was designed for
- give pleasure in use
- have aesthetic appeal
- be safe for its purpose
- be of acceptable quality
- last for a reasonable lifetime
- offer value for money
- cause minimum environmental damage
- have not exploited the workforce in its manufacture.

Manufacturers need to:

- understand and use appropriate production systems
- reduce parts and assembly time, labour and material costs
- apply safe working procedures to make safe products
- test products against specifications before distribution
- produce consistent results by using quality control and quality assurance procedures
- understand and use product distribution systems
- be aware of legislation and consumer rights
- assume legal responsibility
- make a profit on the products produced.

Human factors

People have a wide variety of **physical** and **emotional** needs that have to be taken into account when designing products for them to use. Not only do they vary in height and weight, but in strength and grip, speed and dexterity, etc. Their abilities to see, hear, touch, smell and taste can differ too – senses that are often essential for being able to operate and use a product. Different people also react in different ways to colours, shapes and surfaces – what makes one person feel relaxed and reassured, can make another tense and uncomfortable.

Anthropometrics is the scientific measurement of the physical size and capabilities of the human body.

Ergonomics is the study of the way people behave in, and react to, their living and working environment and the things they use.

Anthropometrics is discussed further on page 92. Ergonomics is discussed further on page 136.

Packaging and labelling

Effective packaging and labelling are essential in ensuring that the consumer can correctly identify that the product they are purchasing is the one they want, will do what they want it to do, and is safe for them to use. It also helps to ensure that when they unpack the item it will not have been damaged in any way.

In many situations the packaging also promotes the product. The colours, textures, images and typography used all suggest the sort of lifestyle the consumer will enjoy if they purchase the product. The graphics of the packaging often communicate the familiar values of the product brand to associate it with the quality of other products made by the same company.

There is more on packaging and labelling on pages 96–104.

Providing the consumer with clear and concise instructions for use is another important part of the design. This is often achieved using diagrams and visual symbols. Sometimes the shapes of the product or its controls help to show how it should be used.

The impact of design

Different people have different needs and wants: what is beneficial or desirable to one person can cause a problem for someone else, or create undesirable damage to the environment.

A new design might enable someone to do something quicker, easier and cheaper, but might cause widespread unemployment or urban decay. It could also have a harmful impact on the delicate balance of nature.

As they develop their ideas, designers often need to make important decisions about the social, moral, cultural and environmental impact of their product.

Social needs

Good design can help bring people together. Designers need to be careful about creating products that might have the effect of isolating someone, or making them more vulnerable to crime in some way.

Cultural awareness

People from different cultures think and behave in different ways. What is acceptable to one culture may be confusing or insulting to another. Colours and certain shapes can have very different meanings across the world.

Moral issues

Sometimes designers are asked to develop products that can cause harm to people or animals. Would you be willing to create a product that could hurt someone or could also be used for criminal activity?

Environmental issues

These are discussed in detail on pages 12–15 and 74–79.

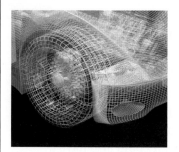

ICT in Product Design: CAD-CAM

CAD-CAM (Computer-aided design/Computer-aided manufacture) is widely used in industry in the development of new products and in their final manufacture.

Computer systems

Computer-aided design (CAD) and **computer-aided manufacture (CAM)** are terms used for a range of different ICT applications that are used to help in the process of designing and making products. **CNC** stands for **computer numerically controlled**.

- CAD is used for creating, modifying and communicating ideas for a product or components of a product.
- CAM is a broad term used when several manufacturing processes are carried out at one time aided by a computer. These may include process control, planning, monitoring and controlling production.
- In a CAD-CAM system the data from a CAD file is sent directly to a CNC enabled machine that automatically makes the component part. In some systems further computer controlled processes assemble some or all of the components of a product.

Computer systems contain three main elements – a series of **inputs** that are transformed into **outputs**.

INPUTS	TRANSFORMATIONS	OUTPUTS
Text, drawings and photographs are put into the system using devices such as keyboards, drawing tablets, cameras and scanners.	The computer's processor and program change the input information as required, providing feedback to the designer via the monitor, and responding to on-screen tools and control devices.	The final design is turned into an output through a variety of devices as instructions to a printer or CNC machine.

Transformation software

A wide variety of different software design applications are available (see pages 68–69). Some of the most common are:

- Pro/DESKTOP and ArtCAM – 2D and 3D modelling
- SpeedStep and TrendStop – textiles design
- CorelDraw, TechSoft 2D, Adobe Photoshop and Illustrator – image manipulation
- Microsoft Publisher and Adobe InDesign – desktop publishing.

Working in 3D

There are three main ways of representing a 3D object on a computer.

- **Wire-frame** modelling means that the object is represented by a series of lines. This image can be enhanced by removing lines which would be hidden.

- **Surface** modelling can then be added. The surfaces of the object are represented by colour, shading and texture to give a stronger sense of the 3D form.

- **Solid** modelling means that the drawing is based on geometric shapes which can be mathematically analysed. These can provide information on such things as the object's mass, volume and centre of gravity.

Rapid prototyping

Pro/DESKTOP drawings can produce stereo lithography files (STL files) which slice the design up into very thin layers. These enable a whole product to be modelled in 3D.

Some rapid prototyping systems use lasers to cure a liquid resin a layer at a time. Other systems extrude a tiny thread of plastic material to build up the design a layer at a time.

This 3D printer sprays a liquid bonding agent on to a tin layer of plaster-based powder, which is then covered with a fresh layer of powder. One major advantage of these systems is that hollow forms can be manufactured. Once removed from the powder, the 3D prototype needs to be dried and treated to make it more durable.

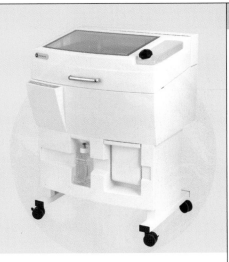

Output hardware

Printers

Graphic work and photos can be printed out at high resolution on to colour **inkjet** or **laser** printers.

Dye sublimation involves printing an image in reverse on to a special paper or plastic film (known as a substrate). Using a heat press, the image is transferred on to the material. Polyester based materials work best for this process. The dye turns into a vapour which then solidifies into the surface of the material, producing a very high quality, durable print.

Special printers are used to print surface patterns directly on to fabrics. CNC embroidery machines can rapidly produce complex designs.

Cutters

Sheet materials can be cut out using a computer-driven **vinyl cutter**. It uses a sharp blade to cut out the net. A **laser cutter** can also be used to cut sheet materials. The laser uses a beam of very intense light to burn through the material. As there is no friction involved, very delicate shapes can be easily cut out.

Millers and routers

A 2.5 axis machine has an X and a Y axis, and a small Z axis. It can be used for engraving and simple milling of sheet metal. A 3 axis machine has an X, Y and Z axis, enabling full 3D shapes to be made. On a 4 axis machine, the work is rotated so that the shape can be cut in one operation.

Typical uses

In industry, ICT and CAD-CAM is used to produce a wide variety of products such as:

- Working drawings, plans and maps.
- Designs for packages, nets and labelling.
- Surface decoration on packaging.
- Books, posters and information sheets.
- Electronic circuit manufacture.
- Fabric and embroidery designs.
- Ceramic surface printing.
- Signs in public spaces.
- Advertising and promotional materials.
- Patterns and moulds for manufacturing processes.
- Quick prototyping.
- Point-of-sale displays.
- Engraving.

CNC knitting and weaving machines are used widely in the textiles industry.

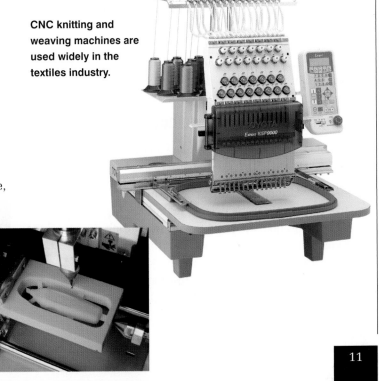

Sustainable Product Design

Industry and consumers can not continue to use up the world's resources. We need to design, manufacture and demand products that minimise the potential damage to the environment.

What is sustainable design?

As things are at present, the western world is using up the earth's natural resources – oil, timber, etc. – at a much faster rate than they can be replaced. Unless something is done, before very long they will simply run out.

At the same time, chemicals and other by-products of manufacturing processes, energy production and vehicle emissions are thought to be causing global warming. This is causing significant changes in temperature of the earth's surface and unsettled weather patterns.

Changing the way we live

Sustainability involves a lot more than sorting out our waste paper and plastic for recycling. It means everyone changing their attitude towards the products they desire and purchase. We need to:

- leave the world in a better condition than we found it
- take no more than we need
- try not to harm life or the environment, and make amends if we do.

Changing the products we buy

Designers and manufacturers have a key role to play. They need to develop products that:

- use fewer materials, fewer components
- use renewable or recyclable materials
- are easy to take apart for repair, upgrading or recycling
- avoid manufacturing processes that produce toxic substances
- minimise the use of energy in production
- reduce the distances involved in the transport of raw materials and finished products.

Natural disasters

Green issues are concerned with the impact a new product has on nature. Some production processes use high levels of non-renewable energy and materials.

Paper and wood
Paper and wood are made from trees. In many parts of the world forests are being destroyed. This is known as deforestation. A high percentage of paper and wood can be used again if recycled, reducing the number of trees that need to be cut down. It is also important to ensure that new trees are grown to replace the ones cut down.

Plastics and metals
Plastics are made from oil and metals are mined from the earth. There are only limited supplies of oil and metal. The amounts of plastics and metals used in a product need to be reduced to the minimum. Some plastics and metals can be recycled.

Fibre and fabric
Synthetic fibres are made from petrochemicals. Dyes and other performance-enhancing finishes often produce toxic waste products that need very careful disposal.

Food
Much of the food we eat is sourced from all over the world. It has to travel a long way to reach our tables. This causes pollution and congestion on the roads.

Decomposition

A product may also be difficult to dispose of when it has been finished with. This can cause further environmental problems. Aluminium can take up to 100 years to decompose. Some materials such as untreated paper and natural fabrics decompose much more quickly. These are known as biodegradable materials.

Who are the bad guys?

How environmentally friendly are the products you use at home every day?

Electronic gadgets
Mobile phones, mp3 players, digital cameras, computers, etc. are made from non-biodegradable plastics and electronic components.

Furniture
Chairs and tables are often made from rare timbers, steel, plastics, treated synthetic fibres and leather.

Clothes
Chemical fertilisers and pesticides are used extensively in cotton production. Fabric dyes create waste products that can pollute rivers. A pair of shoes can use nearly a third of a litre of chemical adhesive.

Lamps
The energy used to light a bulb comes from coal, oil or nuclear fuels, all of which are limited natural resources. Bulbs contain mercury which is toxic, yet they are simply thrown away into a landfill site.

Magazines
Although paper is made from sustainable forests and can be recycled, glossy finishes, printing inks and chemicals are environmentally very unfriendly.

Sustainable Product Design [continued]

When designing products it is important to remember the 3Rs: Reduce, Recycle and Reuse. The 3Rs provide guidance on how to minimise the damage done to the environment by a product.

The recycling symbol is based on the idea of a mobius strip, which forms a continuous surface of material. It was designed in 1970 by Gary Anderson, a 23-year old student.

The challenge

Many consumers already recycle goods and packaging. Now it's up to designers and manufacturers to ensure they reduce the amount of materials and energy used in production, and to reuse materials and components.

What are the 3Rs?

Reduce
Designers and manufacturers need to aim to use the least amounts of materials and energy in making a product.

Recycle
Designers and manufacturers need to use recycled materials and/or materials that can be recycled after use (e.g. untreated papers, glass, some plastics). Recycled materials are those that can be used again in new products. This usually involves separating the materials into different types and then cleaning and re-preparing them.

Reuse
Designers and manufacturers need to use materials and components (e.g. containers, electronics, fastenings) that have already been used and/or can be used again in different products. This can sometimes be easier and cheaper than recycling. Designing with this in mind involves making sure that a product can be easily disassembled after it has been discarded. It is also necessary to check that surface finishes will not degrade the underlying material.

Other approaches are to:

- make more products **upgradeable, multi-purpose** and/or **durable** and **repairable**
- provide more products that are **sharable** by a local community or cooperative
- source materials **locally** to reduce transport costs.

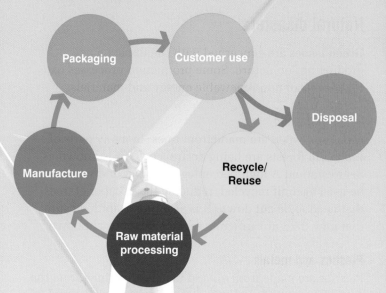

From Cradle to Grave: developing a product lifecycle details the impact it will have on the environment, from raw material to the end of its useful life and beyond.

Sustainable graphics and packaging

The environmental aspects of **packaging** are of particular concern. Here are some of the things that designers and manufacturers need to do:

- Use materials such as recyclable paper and card instead of plastics.
- Minimise the amount of materials used, particularly where the packaging will only be used once and then thrown away.
- Reduce the size, shape and weight of the packaging to minimise transport costs.
- Provide refillable containers.
- Avoid packaging materials that have been chemically treated during their production (e.g. using CFC gases), or that will poison the ground when decomposing.
- Reduce the number of separate parts to make the package easier to throw away.

Sustainable 3D products

At the same time as making products functional, attractive and desirable, designers need to:

- ensure that raw materials such as timber are replenished
- plan for what happens to a product when it is discarded
- reduce the number of components and different manufacturing processes.

Sustainable textile products

To reduce the environmental impact of textile products, designers need to:

- include more natural fibres and dyes
- use recycled or reused materials and components wherever possible
- reduce the amount of energy needed for manufacture
- avoid the use of toxic chemicals in dying and finishing processes.

Sustainable food products

In terms of the production and distribution of food products, designers need to:

- specify the use of locally-grown organic ingredients
- reduce the amount of energy needed for processing
- promote the use of fair trade produce
- reduce the number of chemical additives while maintaining appearance
- minimise the amount of packaging required.

What's the score?

Here's a quick way to obtain a rough comparison about the sustainability of different products. The lower the score, the more sustainable a product is.

- Is it made using recycled cardboard, ceramics, wood, untreated fibre, and/or naturally-grown food ingredients? If so, give it a score of **1 mark** for each material used.
- Is it made using glass, leather, plastic, paper, rubber, steel, manufactured fabric or processed food ingredients? If so, give it a score of **5 marks** for each material used.
- Is it made using aluminium, electronic components, light bulbs, paint, polystyrene and/or stainless steel? If so, give it **15 marks** for each material used.
- Is it made using batteries, brass, chromed steel, copper, lead and/or zinc? If so, give it **50 marks** for each material used.
- Estimate the number of types of different component parts. Add **5 marks** for each different type of part.
- If you think it was probably made locally, add **1 mark** for transport. If you think it was probably made in this country, add **10 marks** for transport. If you think it was probably made in another country, add **20 marks** for transport.

How does it score?

1–10 marks	Very good.
11–25 marks	Quite good.
26–50 marks	Needs improvement.
More than 50 marks	Unsustainable.

This Remarkable Pencil™ is made from one recycled plastic cup

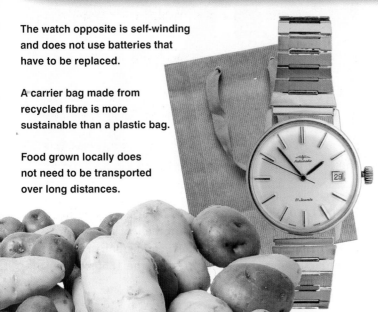

The watch opposite is self-winding and does not use batteries that have to be replaced.

A carrier bag made from recycled fibre is more sustainable than a plastic bag.

Food grown locally does not need to be transported over long distances.

Environmental symbols

Graphic symbols are used to show the environmental make-up of a product and its packaging. Some show the recycling processes involved. Others help remind users to dispose of the product and its packaging in the most effective way.

These symbols are often coloured green.

The symbol on the left shows the percentage of recycled materials used.

The symbol on the right is used across Europe to show the use of environmentally friendly packaging.

A brief history of Product Design

New designs do not simply happen on their own. They evolve out of the available technology of the time, people's changing needs and lifestyles, and the social, economic and political circumstances of the time.

Although there have been many product designers, very few become well known. Those who do have influenced major styles and trends in a wide range of familiar everyday products.

The age of artefacts

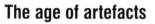

The first things to be designed were simple artefacts, usually made by the user for themselves. These were probably things such as arrowheads, simple cutting tools, ceramic pots and roughly woven garments. Over thousands of years of evolution, such items came to be decorated and jewellery, headdresses and religious icons started to be made.

Cave paintings and hieroglyphics (a system of visual symbols) were amongst the earliest forms of recorded communication. During the eighth and twelfth centuries, the Chinese printed text and pictures using hand-carved wooden blocks. In Europe, monks produced hand-drawn illuminated manuscripts, but these could only be afforded by the very wealthy.

Everyday cloth was produced locally but an international market in luxury textiles developed. The movement of textiles within Europe was extensive and trade even reached to the Far East. During the first century BC, Chinese textiles brought by camel caravans reached the Mediterranean.

Making things in quantity: the Machine Age

Graphic items were some of the first products to be made in quantity. In the fifteenth century in Germany, **Johanns Gutenberg** invented the movable type printing press. This made printing much faster and cheaper, so that many copies could be printed and distributed.

The development of steam power in the seventeenth century led to the Industrial Revolution. Giant machines could now produce textiles, tools and objects such as ceramic and metal containers in large quantities, though these were essentially simple and functional, and neither as well made nor as attractive as hand-made items. By the time of the Great Exhibition in 1851 (held at Hyde Park in the Crystal Palace, designed by **Joseph Paxton**), items had become over-decorated and elaborate, often to hide poor workmanship.

By the 1850s the replacement of natural dyes with synthetics was well underway in the textiles industry. The sewing machine for domestic and small-scale manufacture was introduced during the 1850s.

William Morris

William Morris (1834–1896) was the founder of the Arts and Crafts Movement. He was opposed to the poor quality of mass-produced goods of the time and the pollution of the environment caused by industrialisation.

Morris was also a talented designer of such products as furniture, wallpaper and carpets, as well as being a leading writer of prose and poetry.

Hector Guimard

Hector Guimard (1867–1942) is most famous for his designs for the Paris Metro at the turn of the century. Clearly based on Art Nouveau-style curvilinear forms and organic shapes, the station entrances he created were uniquely based on standardised mass-produced

cast iron component parts that could be specified according to the requirements of the location.

The Product Age

The concept of Product Design, and the product designer, only emerged during the latter part of the nineteenth century. For the first time, neither user, maker nor manufacturer determined the nature of the objects being made; but someone else, who had knowledge of what people wanted and needed and of the materials and processes that could be used to make useful, satisfying and attractive 'products'.

Design movements

Since then, the history of Product Design can be traced through a series of 'movements'. These refer to particular approaches to design, mainly in terms of the shapes, forms, colours and textures of products in response to:

- the social and economic circumstances of the time
- the possibilities of new developments in material and component technologies and mass-production processes.

Although the various design movements often originated in one country and were led by one or two innovative designers, they spread rapidly across Europe and America, with each country developing its own application of the approach.

It's easy to think that people's homes were filled with things that were in the style of the time. The reality is that only a few people could afford the latest fashions. However, the designs of some household products were influenced by these 'movements' and had some of their basic characteristics.

Arts and Crafts

The first 'movement' in design is generally accepted to be the **Arts and Crafts Movement**. This was founded in the 1890s by William Morris with the aim of improving the quality of craftsmanship in products and the conditions of the workers who produced them. The belief was that design should be based on simple, organic forms derived from nature. The style is characterised by natural shapes and high-quality materials. Unfortunately, the cost of these products was high, so only the wealthy could afford them.

Art Nouveau

The **Art Nouveau** style developed across Europe (especially in France and Belgium) and America during the late nineteenth and early twentieth centuries, but did not become widespread in England. It is characterised by the use of organic, free-flowing lines and shapes. Examples can be found in architecture, wrought ironwork, glass and furniture, jewellery, fabrics, wallpaper and advertising posters where the design of the flowing, organic typefaces were interwoven into floral and decorative devices and swirling figures. The Art Nouveau designs of each European country and in the USA all have subtle differences.

A brief history of Product Design [continued]

The Modernist Movement

The industrial designs which came to characterise the twentieth century originated towards the end of its first decade. The designs of **C.R. Mackintosh** and the Viennese designers **J.M. Olbrich** and **Otto Wagner** began to use geometric shapes which had more potential for mass production. In 1907, the German **Peter Behrens** was appointed design coordinator to the electrical firm AEG, and was responsible for the design of the factory buildings, its products and publicity material. The logo he designed probably represents the first example of the implementation of a corporate identity scheme.

Dutch, Italian and German designers began to revolutionise typography and graphic layout. They sought to create images and forms that were appropriate to the Machine Age. Words became abstract shapes, and the symmetry and decoration of Victorian times became a thing of the past.

The Bauhaus

These new industrial styles were further developed and refined by the designers of the **Bauhaus** between 1919 and 1933. The Bauhaus was a school of art and design founded in Germany by **Walter Gropius**, who had earlier worked for Peter Behrens. It laid down many of the principles of design that are followed today, and that are still taught in schools and colleges. One of the aims was to produce designs suitable for mass production, though it was not until the latter part of the century that their designs for products such as furniture and lighting truly came to be made cheaply and in quantity. Another aim was to produce designs that were 'international', i.e. that would have appeal across all western countries.

Many of its students went on to become leading designers in the fields of architecture, furniture design, ceramics, metalwork, textiles, stage design and photography. Many famous European artists also studied at the Bauhaus.

Have you heard of... ?

Charles Rennie Mackintosh

C.R. Mackintosh (1868–1928) was a Scottish architect, interior and furniture designer, and artist. Although generally recognised as part of the Art Nouveau Movement, the sparseness and purity of his interiors and the use of more geometric forms anticipated the 'Modernist' style which followed in the 1920s and 1930s. Like other innovative designers of the time, he and his wife (Margaret Macdonald Mackintosh) aimed to integrate architectural and decorative elements throughout the buildings they created.

Henry Ford

Henry Ford (1863–1947) founded the Ford Motor Company in the USA. He was one of the first to develop mass-production assembly lines in order to produce affordable motor cars. The famous 'Model T' was introduced in 1908. By 1918, half of all cars in America were Model Ts.

Art Deco

From 1925 onwards, at the same time that the work of the Bauhaus was in full development but not yet widely applied, the **Art Deco** style became extremely popular and influential. It was named after an international exhibition in Paris in 1925 in which all exhibits were required to be novel in their design.

The style is characterised by many visual elements, such as bright colours, images of the sun, and geometric shapes and zigzag patterns derived from Egypt, inspired by the discovery of Tutankhamen's tomb in 1922. There was also an emphasis on the use of unusual, exotic and expensive materials.

During this time, there was a considerable demand for eye-catching display typefaces for use on posters and packaging, and these often reflected the Art Deco style and the early work of the Bauhaus.

The ideal home

It was not until the 1930s that most people began to acquire 'designed' products in any quantity. The increasing cost of the domestic servant and the growth of the national electricity grid across the country led to a demand for the first electrically-powered household products such as radios, vacuum cleaners, fires and washing machines.

There was a great interest in speed at the time, with land, sea and air records being broken regularly. As a result, many household products were 'streamlined', an effect that had become easier to achieve with new production technologies.

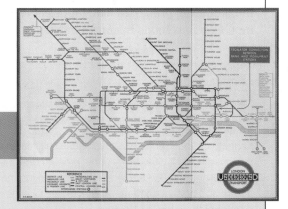

Coco Chanel

Coco Chanel (1883–1971) was a pioneering French couturier who had a major influence on fashion design in the twentieth century. She is famous for her designs in the early 1920s of a blue sailor skirt and masculine garments for women working in factories and offices. She was the first designer to use wool for women's suits, which was cheaper than silk. Chanel also pioneered knitted fabrics and the wearing of 'costume' jewellery. The Chanel Number 5 fragrance was launched in 1921, and was one of the first to establish the relationship between fashionable clothing and perfume.

Harry Beck

Harry Beck (1903–1974) designed the innovative schematic 1931 London Underground map, still in use today. Inspired by electric circuit diagrams, the routes were organised along vertical, horizontal or 45-degree axes, and also enlarged the distance between the stations in the central area, and reduced the distances in the outer area. As well as being easier to understand, it suggests that long journeys are shorter than they actually are, and that the interchanges between lines at some stations are much closer than in reality: this led to growth in passenger usage. The basic concept has been copied by metro systems across the world.

A brief history of Product Design [continued]

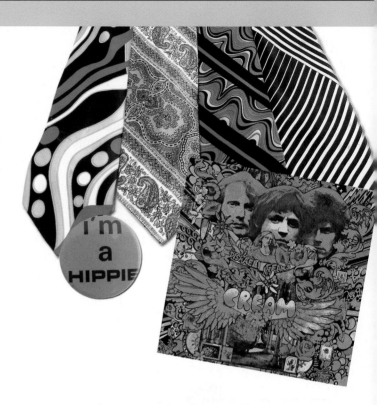

Post-war products

During the Second World War, significant advances were made in materials, electronics and user-ergonomics. However, the cost of post-war reconstruction meant that rationing of food, furniture and clothing continued into the early 1950s, with the result that the consumer was keen for exciting bright new products. In the 1950s and 1960s there was a consumer boom. 'Youth culture' began to provide a distinctive new market, keen to establish its own identity and sense of individualism.

The first modern plastics such as foam, nylon and polyester were developed. Plastic materials revolutionised industrial design and production, as the moulding processes made it possible for the first time to create goods in considerable quantity at low cost.

During the 1960s, the earlier post-war interest in Science gave way to images led by the Space Race and the first moon landings, science fiction, fantasy, and 'futuristic' designs. These were widely applied to household products that were becoming increasingly automated. Late 1960s graphic products used 'psychedelic' colours and patterns, and decorative typefaces and designs that often referred back to Art Nouveau styles.

Have you heard of... ?

Kenneth Grange

Kenneth Grange (born 1929) worked initially as an exhibition designer in the 1950s. Many of his product designs from the 1960s are familiar today. He designed the first parking meters, food mixers and pocket cameras. In the 1970s, Grange designed the exterior of the InterCity 125 train.

Grange's successful innovations resulted not just from an interest in style and the use of new plastic materials and processes, but from his consideration of the function and use of the products.

Mary Quant

Mary Quant (born 1934) is credited with the invention of the mini-skirt in the 'Swinging Sixties'. She opened the first boutique – a small, overcrowded fashion shop – called 'Bazaar' in London in 1955. Quant was one of the first fashion designers to create clothes and accessories aimed at young people at affordable prices. She also used young, unsophisticated models, such as Twiggy, to model her designs. Her clothes sold in large quantities in the UK and internationally.

Have you heard of… ?

Alberto Alessi

Alessi is a family-run Italian company, founded in 1921. Many of the Memphis designers worked for Alessi at some point during the 1980s and 1990s, including Ettore Sottsass, Michael Graves and Philippe Starke. Their kettles and corkscrews, coffeemakers and condiment sets continue to lead the way in the design of kitchen and dining products.

Neville Brody

Neville Brody (born 1957) is currently one of the UK's best-known graphic designers. He designed record covers before becoming a magazine art editor. Brody's innovative work in the early 1980s for the 'Face' magazine was highly influential. He began to develop and use typefaces as graphic devices, and experimented with unusual surface textures and decorative motifs that often ignored the conventions of layout grids and typesetting. Brody's work was often imitated during the late 1980s in packaging, magazines and record covers.

James Dyson

James Dyson (born 1947) invented the Cyclone bag-less vacuum cleaner back in the 1970s but failed to persuade any established manufacturers to invest in it. In the 1990s he set up his own manufacturing company, and his cleaner is one of the most popular brands in the UK. Dyson is not just a designer, but an inventor, engineer and entrepreneur.

Into the 1980s: the Marketing Age

The 1980s and 1990s represented a significant shift from the product itself towards the creation of the global market for a product and the lifestyle brand image that will persuade people to buy it. The promotion and packaging became all-important.

Previously, the name of the 'designer' of a product was of no interest or importance. Now, for the first time, the idea that the product had been 'designed' by someone famous added value, and 'celebrity designer' jeans were quickly followed by other clothes and fashion accessories.

Meanwhile, designers were free to draw on colours, shapes and forms from a rich variety of past styles and from countries across the world.

Another theme of the last quarter of the twentieth century was increased miniaturisation, led by developments in micro-electronics and new materials technologies. Computer-aided design and manufacture began to revolutionise the production industries, particularly in terms of plastics and fabrics.

The Memphis Group

The **Memphis Group** was a highly-influential Italian design movement of the 1980s founded by **Ettore Sottsass**. Although their work was centred around furniture, architecture and textiles, their impact on the public is more apparent today in graphics and homeware. The group's work was a reaction against the brutal 'minimal' designs that preceded it. They mixed styles, colours and materials with a sense of humour.

The design drawings of the Memphis Group are of particular interest. They are often simplistic and child-like and, as such, in contrast to the more formalised and realistic sketches of the industrial designers of the 1950s, 1960s and 1970s.

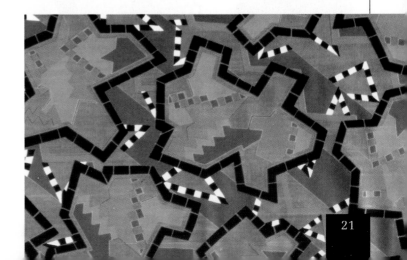

A brief history of Product Design [continued]

Twenty-first century Product Design: the rise of the blobject and the gizmo

At the start of the twenty-first century the nature of Product Design is changing again.

The need for sophisticated 3D animation software in the movie industry has led to the development of programs that can push and pull, stretch and fold complex curves or shapes without the need to construct mathematical structures.

Blobjects are objects with a curvilinear, flowing design, such as the Apple iMac computer and the Volkswagen Beetle. A blobject is a physical organic, colourful blob-shaped object that has been redesigned or created entirely on a computer graphics screen and shaped with a mouse. It has then usually been injection moulded in plastic, or made in metal or rubber. Buildings that have been designed this way are known as **Blobitechture**.

At the same time, another type of product has begun to emerge – the **gizmo**. Gizmos are small multi-functional electronic devices that might include a phone, web browser, camera, keyboard and word-processor, notepad, sketchpad, calendar, diary, clock and mp3 player. Even more accessories can be plugged in. Unlike a machine or a more traditional product, gizmos have a short lifespan, as new model versions with even more technical functions crammed in appear every six months. At present, most gizmos are a long way from being attractive, user-friendly products.

Selfridges, Birmingham, designed by Future Systems.

The 'Gherkin', London, designed by Norman Foster.

An iPAQ Personal Digital Assistant.

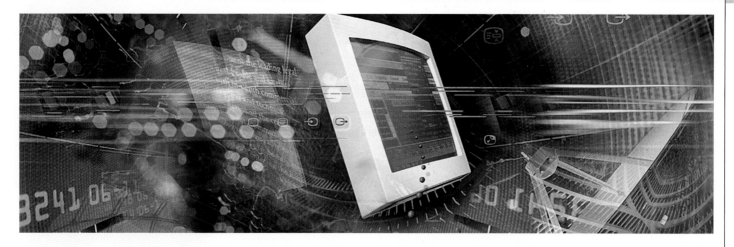

The future of products in the Information Age

During the early twenty-first century, developments in electronic communication technologies will continue to develop rapidly in new and exciting ways. The future of Product Design lies in a further convergence of electronics and communication devices, new 'smart' materials that respond to their environments, nano and bio-technologies, and personalised target marketing.

Meanwhile products also need to become more sustainable though the use of renewable resources and small-scale local manufacture using CAD-CAM technologies that meet local rather than global needs.

Tomorrow's world

In the not too distant future, when you buy an electronically tagged product, you will easily be able to discover:

- where it was, when you got it and how much it cost
- who made it, what it was made of, and where those resources came from
- who to thank for making it and who to complain to about its inadequacies
- what previous kinds of objects used to look like and why this one is better than earlier ones
- what people think the next version might look like, and what you could do to help that happen
- the history of its ownership, what it had been used for, where and when it was used
- what other people who own this kind of object think about it
- how other people more or less like you have altered or modified their object
- what most people use their objects for, and how much it is worth on an auction site
- and especially, absolutely critically, where and how to dispose of it safely and responsibly.

Anything, any time, any place

Tomorrow's products could be very different in their nature, and probably won't even be called products! All objects will be electronically tagged, and information about them recorded and kept digitally. The point of origin in time and space of each component and surface will be recorded. Any changes to its form and location, and the frequency with which it is used will be known. Each product will each have its own history. Our purchasing interests and habits will become even more valuable data in the drive to create new things we need and desire.

Information about information

The most valuable commodity of the twenty-first century will not be precious artefacts, but information itself. There will be so much data stored about everything and everyone, that information about where the information is and how to find it will be even more valuable. But the thing that will be most sought after is likely to be privacy – the ability to live without leaving a trail of information about ourselves.

Properties of materials

Different materials behave in different ways. Some products require materials that have certain properties, while others need a completely different combination of properties. Identifying the necessary properties and choosing the right materials is an important part of designing.

What are properties and characteristics?

When designing a product, the **physical properties** and **working characteristics** of the materials are very important. The designer needs to know about their:

- properties – how materials perform in everyday use
- characteristics – cutting, joining and finishing them.

 Pages 24–29 are mainly concerned with the properties of materials.

Behaving properly

Different materials behave in different ways.

Some resist having their shape changed, while others are easy to bend and fold.

Some allow electricity and heat to pass through, while others are poor conductors.

Some change in a predictable way in response to heat, light or moisture.

Some are more environmentally friendly to produce and/or dispose of than others.

Some soak up liquids quickly, while others repel them.

Some are cheap to manufacture and process, others can be costly.

Some are heavy, while others can be very light.

Some scratch easily, while others are difficult to mark.

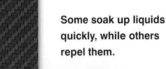 Some snap in two very easily, while others can be difficult to separate.

Properties of resistant materials

There are eight common properties for all resistant materials:

1 Stiffness
Stiff materials are not flexible; they cannot be squeezed or stretched easily. Glass will shatter if it is squeezed or stretched, for example.

2 Strength / ductility
Strong materials are not easy to break or deform. So, a steel girder can support a heavy load without bending.

3 Toughness
Tough materials absorb a lot of impact before they break. A wooden mallet is used to strike the end of a chisel. This means a lot of impact each time the mallet is used. But the mallet can last the lifetime of the user without breaking.

4 Thermal conductivity
Thermal conductivity is how easily heat is passed through the material. Metals are good thermal conductors, for example a steel kettle gets very hot when water is boiled in it.

5 Electrical conductivity
Electrical conductivity is how easily electricity is passed through the material. The core of an electric cable is made from copper because copper is a good conductor of electricity.

6 Hardness
This is how resistant to wear the material is. Floors in schools are often made from wood because, although they are walked across every day by many thousands of feet, wooden floors take a long time to wear out.

7 Density
This refers to how heavy a material is when compared to its volume. Low-density materials, such as expanded polystyrene, are light for their volume. The metal lead is heavy for its volume – it is a high-density material.

8 Environmental resistance
How easily the material rots or corrodes in sunshine and rain, or how resistant it is to fire. Outdoor furniture is often made from teak, because this wood is resistant to rot.

Property sharing

Metals, woods and plastics all share these eight properties. However, different materials display different combinations of these properties. For example, metal is generally tougher than plastic, but plastic has a better environmental resistance than wood.

Sometimes a design requires a material that is high in one or more particular property, such as toughness. However, it might also require a material that is poor at doing something, such as electrical conductivity. For example, the casing of a torch is usually plastic or rubber, both of which have low electrical conductivity.

This table shows the ways in which metal, wood and plastic vary in their properties.

Property	Metal	Wood	Plastic
Stiffness	High	Medium	Low
Strength	High	Medium	Low
Toughness	High	Low	Low
Thermal conductivity	High	Low	Medium
Electrical conductivity	High	Low	Low
Hardness	Medium	Low	Low
Density	High	Low	Low
Environmental resistance	High	Low	High

There are also differences between different types of the same materials. For example, some metals are stronger than others and some plastics are stiffer than others.

 See also 'Working smarter' on pages 140–141.

■ TEST YOURSELF

1. A designer is developing ideas for a new picnic table for use in local parks. What sort of properties will the chosen materials need to have? **(6 marks)**

2. Choose *three* different products illustrated in this book. List the materials used in each and give a reason for their use in terms of the required properties. **(6 marks)**

Properties of materials [continued]

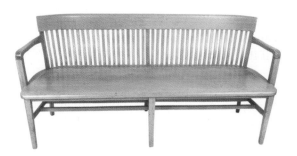

Metal matters

Designers make choices about which metals they are going to use by considering the properties of the metal and whether these suit the needs of their design. For example:

- Does it need to be light or heavy?
- Does it need to be malleable, flexible or rigid?
- Does it need to resist corrosion?
- What colour is it?

- Is it shiny or dull?
- Is it tough?
- Is it environmentally friendly?
- Is it a good conductor of heat?
- Is it a good conductor of electricity?

- Would an alloy of metals provide a good solution?
- How much does it cost?
- How can it be finished?
- How can it be joined?
- What manufacturing processes are needed to shape it?

Ferrous or non-ferrous?

Metals are categorised as ferrous or non-ferrous. This refers to whether they contain iron or not. For example, mild steel is a ferrous metal – it contains iron and 0.15–0.3% carbon. Adding carbon gives it different properties from pure iron, making it tough and malleable.

Aluminium is a non-ferrous metal – it does not contain any iron. It is made from the pure metal with no additions.

Alloys

Alloys are combinations of metals. For example, brass is made from 65% copper and 35% zinc. The resulting alloy combines the properties of both metals.

Heat treatment

When metal is heated and left to cool slowly the process is called annealing, which makes it more malleable and ductile. An oxide is formed on the surface but can usually be removed by soaking in alum solution or pickling.

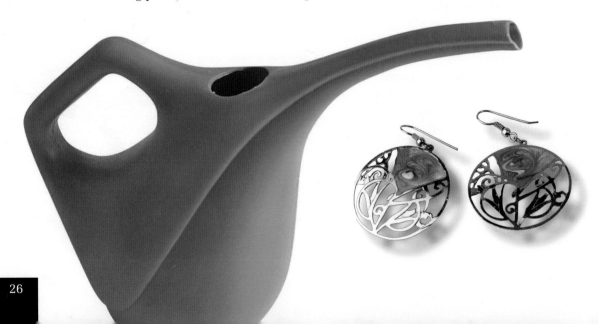

Wood – you know it!

There are two categories of material that are often referred to as wood: natural wood and manufactured boards. Designers make choices about which natural wood or manufactured board they are going to use by considering whether the properties suit the needs of their design. For example:

- Does it need to be light or heavy?
- Does it need to be straight grained?
- Does it need to be free from knots?
- Does it need to be weatherproof?

- Does it come in wide boards?
- What colour is it?
- Can it be veneered or laminated?
- Can it be bent in a curve?
- Is it environmentally friendly?

- How much does it cost?
- How can it be finished?
- How can it be joined?
- Is it easily worked?

For more about the difference between natural wood and manufactured boards see page 75.

Plastic properties

Designers make similar choices about the working properties of the different sorts of plastics they are going to specify.

Plastics are categorised as **thermoplastics** or **thermosetting plastics**. This is a good guide to the working characteristics of the plastic.

Thermoplastics

Thermoplastics are shaped when they are hot. They are less resistant to heat and fire than thermosetting plastics. Thermoplastics can be heated after being formed. They return to their original flat sheet form. This is known as **plastic memory**.

An example of a thermoplastic often found in school is acrylic – a familiar trade name for this is Perspex™.

Thermosetting plastics

Thermosetting plastics are generally stronger, harder and stiffer than thermoplastics. They are more resistant to heat and fire. A chemical reaction is involved in their forming.

Polyester resin is an example of a thermosetting plastic. It is laminated with glass or carbon fibres for moulding GRP (glass reinforced plastic) car and boat bodies.

■ **TEST YOURSELF**

1. **a)** A designer is working on an idea that combines metal and wood in a design for a chair. Which metal would be most suitable for constructing the chair legs? **(1 mark)**
   b) Write about *four* properties that this metal has which makes it suitable for this use. **(4 marks)**

2. A manufacturer wants to produce a new range of kitchen bins. He has approached you, as a designer, with the idea of using high-density polythene.

   What *four* properties does this plastic have that makes it an ideal choice? Give your reason for each choice. **(8 marks)**

3. Study the products on these two pages. Suggest what materials each could be made from. Explain what particular properties these materials would need to have to be suitable for the product. **(6 marks)**

Properties of materials [continued]

Choosing the right fabric

The designer needs to consider:

- who the product is for
- the environment it will be used in
- the intended price range of the product.

A good knowledge and understanding of different fibres and fabrics is needed in order to be able to judge their fitness for purpose for a particular end-use.

The structure of a fabric and whether any finishes have been applied to it also need to be considered. It is also important to be aware of the techniques and processes used to manufacture the intended product, as these can vary according to the characteristics of the fabric being used.

About fibres

Some fibres are more **resistant** to **abrasion** than others, for example, nylon and polyester. Some fibres are highly **absorbent**, such as silk, cotton and wool. While others, such as acrylic, allow very little **moisture** to penetrate.

Silk and wool have very good **flame resistance**, while linen, cotton and acrylic are highly flammable. Silk and wool are also good **insulators**, while linen and cotton are not. Nylon and polyester have high tensile strength, while wool does not.

Reference is often made to a **fibre property chart**, which provides a good general guide to fabric properties and characteristics. Fabric testing then ensures that the fabric has the right qualities for a product.

Fitness for purpose

It is important to choose fabrics that are fit for their purpose. In industry a **fabric specification** is produced. This sets out the essential and desirable criteria for a fabric, taking into consideration its intended use.

For example, the main considerations for a luxury soft furnishings fabric might be aesthetic qualities and stain resistance. Therefore, a rich velvet or brocade with a stain-proof finish might be a good choice.

For work overalls, durability and ease of laundering might be more important. A strong polyester/cotton might be a good choice for these.

For children's toys, safety and non-flammability will be very important when determining the fitness for purpose of a fibre or fabric.

	natural fibres				man-made fibres				
	Silk	Wool	Linen	Cotton	Acetate	Acrylic	Nylon	Polyester	Viscose
Abrasion resistance	**	**	**	**	*	***	****	***	*
Absorbency	****	***	***	***	*	*	*	*	**
Elasticity	**	***			**	***	***	***	**
Flame resistance	***	***	*	*			***	**	*
Insulation	***	****			*	***	***	***	
Mothproof	*		***	***	***	***	***	***	***
Mildew resistance	*	*			***	***	***	***	***
Resistance to acids	*	***	*	*	***	***	***	***	***
Resistance to alkalis									

Working well together

Different fabrics can be combined together to improve their properties and performance characteristics. The balance between properties and characteristics, such as permeability, fabric handling and appearance, can be chosen to suit a particular application. This may also create a safer fabric.

- Interfacing can be combined with a fabric to produce strength and stiffness.
- Quilting different fabrics in layers can provide added warmth.
- Combining fabrics can create decorative effects, as well as serving a useful function. For example, padding and quilting can create texture as well as adding warmth to a garment or quilt. These techniques involve sewing fabrics together.

Padding

Appliqué work can be padded by inserting a piece of wadding behind the fabric pieces. This creates a 3D effect and can give your work added depth.

Quilting

This involves placing wadding between two layers of fabric and stitching on top to create a pattern. This is one of the best ways to insulate an anorak or bedcover. Polyester wadding is designed to trap as much air as possible within the structure of the fabric and more air is trapped between the three layers of fabric.

Coating and laminating

Fabrics can also be combined, using polymer film to either coat or laminate.

Coating is when a layer, or layers, of a polymer film is applied to the surface of a base fabric. These fabrics may keep the wearer dry in a storm, while still allowing perspiration from the body to escape. These are known as breathable fabrics.

Laminating is when two or more layers of fabric are combined by an intermediate layer of polymer which adheres to both surfaces. Such fabrics allow the attractive appearance of an unstable knitted or lace fabric to be combined with the stability of a woven fabric.

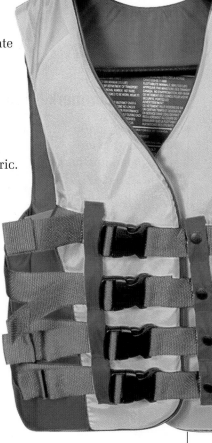

■ TEST YOURSELF

1. Identify *two* properties of cotton and explain why cotton is suitable for garments such as T-shirts and underwear.

   (4 marks)

2. Identify *one* property of nylon that makes it popular for rucksacks. **(2 marks)**

3. Study the products on these two pages. For each, suggest what materials and/or fabrics they might have been made from. For each product write a list of the properties the materials and/or fabrics would need to have. **(9 marks)**

Marking and cutting out

Before any material is cut out by hand it first needs to be marked out. Then, careful choices need to be made about which is the best tool to use to cut the material.

Making your mark

Depending on the materials you are using, marking them out is likely to require tools such as:

- a tri-square
- an engineer's square and scriber
- a sliding bevel
- a combination square
- a marking gauge
- a mortise gauge
- a marking knife
- measuring tools.

Hold tight!

Materials need to be securely held while they are being cut or worked on in some way. Again, depending on the specific type of material you are using, you might need to use:

- a woodworking vise
- an engineer's metal-working vise
- a workmate bench
- a G cramp
- pliers
- a bench hook
- a mitre box
- sash and mitre cramps (or clamps).

Cutting it out

Saw points

Saws are used for cutting resistant materials to size. The one you choose depends on the particular material and the length and depth and accuracy of cut required. Which of the following saws will you need?

- Hand saw.
- Tenon saw.
- Dovetail saw.
- Coping saw.
- Fret saw.
- Hacksaw.
- Junior hacksaw.

There are also a range of machine and power saws:

- Circular saw: fitted with rip or cross-cut blades for wood.
- Powered hacksaw: for cutting steel up to 150mm diameter.
- Band saw: different blades for wood, metal or plastic.
- Electric jig saw: uses a variety of blades for wood, metal or plastic.

You will probably need special permission to use these tools.

Drilling holes

When it is inappropriate to use a power tool, a hand drill or bit and brace may be used. For example, where the size of the chuck is smaller than the size of bit required or the location of the proposed hole makes the use of a power tool dangerous.

There are a range of different types and sizes and diameters of drill bits. The correct one needs to be chosen according to the size and depth of the hole, the material and the drill being used.

Wood you believe it!

There are a range of different forming, cutting, joining and finishing processes involved in working with wood and manufactured boards. You might need to choose an appropriate one:

- Chisel – bevel, firmer or mortise.
- Plane – jack or smoothing.
- Lathe – for turning.

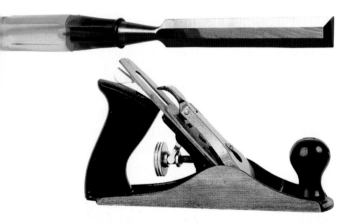

Joints

Different joints are used for different purposes:

- Mitre and butt joints.
- Mortice and tenon joints.
- Dovetail and comb joints.

Laminating

Thin sheets of material (wood and plastic) are built up in layers with adhesive and bent to the required shape. Formers are used to support the curves until the adhesive dries. The results are very strong and used  for tables, chairs and even bridges!

Milling machine: metal and plastic

Milling machines are used to cut horizontal and vertical surfaces and slots. There are two types of milling machine: horizontal and vertical, depending on where the cutters are mounted. Horizontal machines use cutters that are wheel-shaped. Vertical machines use cutters that look like drills. The table moves in all three axes (x, y and z).

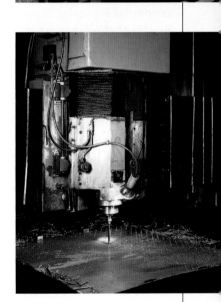

Centre lathe: metal, plastic and wood

This machine cuts very accurately and uses tools that are held in a tool post that can be moved in two directions.

Strip heater: plastic

A strip heater is used for heating thermoplastic sheet so that it can be bent.

Vacuum former: plastic

This machine is used to heat thermoplastic sheet so that it can be sucked over a mould or former to create a 3D shape. You may have access to other specialist tools, such as an injection-moulding machine, or CAM equipment.

Fixing it all together

There are a wide variety of pre-made components and adhesives that are available for joining different materials together temporarily or permanently.

Deciding on the way in which similar and dissimilar materials are joined is an important part of the process of designing. Ideally, there should be as few fastening operations as possible in order to reduce production time and costs. Durability, safety, cost and ease of working also need to be taken into consideration.

Nuts and machine screws

Nuts and machine screws (bolts) are removable fastenings for metal, wood and plastic. They are usually used with a washer, which is placed between the nut and the machine screw.

Types of nuts
- Hexagon: the basic nut, tightened with a spanner.
- Wing nut: used with removable fittings, tightened by hand.
- Locknut 1: used with a standard nut to stop the first nut from coming loose. The two nuts are tightened against each other.
- Locknut 2: nylock-type nuts have a nylon insert and resist coming loose without a second nut being used.
- Cap nut: this gives a decorative finish at the end of a bolt.

Types of machine screws

There are two main types of machine screws:

- A standard machine screw is only threaded for part of its length.
- Set machine screws are bolts that are threaded for their whole length. These are screwed into a threaded hole to fix two parts together.

Both types come in a variety of lengths and diameters.

There are many types of machine screw heads, including slot heads, countersink and cheese head. Hexagonal or socket head screws are known as Allen screws.

Sizes of machine screws

Machine screws are measured by the length of the body, not including the head. 'M' means the outside thread diameter in mm. The thread is either coarse or fine.

Washers

Washers are available in different ranges of sizes and thicknesses. Their size is the diameter of the hole measured in mm. There are several types of washers:

- Plain washer: used to spread pressure over a larger area and protect the surface of the part being joined.
- Spring lock washer: stops the nut from coming loose. Made from sprung steel.
- Toothed washers: these have spikes that bite into soft material, providing extra grip.

Screws

Screw heads

These come in a variety of types. Decorative press covers are available for Phillips and Pozidrive screws. Dome head screws have a metal dome and are usually used for mirror fixing. Countersunk headed screws need a countersunk hole for a flush surface finish.

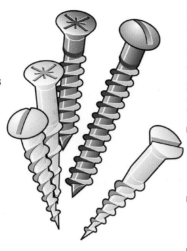

Sizes of screws

The size of a screw is the diameter of the shank, or unthreaded part. Screw sizes are measured by gauge numbers. These go from 4 to 12 in even numbers. 4 is small and 12 is large.

The length of screws is measured from beneath the head to the tip. Each gauge of screw comes in a range of lengths, varying from 10mm to 100mm.

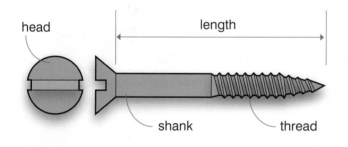

Fastening wood

Wood is often joined using screws made from steel, brass, aluminium and stainless steel. They are available in a variety of different finishes including: bright zinc plate; nickel plate; chrome plate; brass plate; and black japanned.

- The heads of steel screws should be painted with metal primer if they are used in damp conditions as they corrode. Non-corroding steel screws are available.
- Brass screws are used with decorative brass fittings. They are expensive and break easily. It is important that the correct size of pilot hole is drilled first.
- Aluminium screws are useful for wood as they do not stain or rust.

Non-corroding screws are also available.

Fastening metal and plastic

Self tapping screws are used to join thin sheet material together, or fittings to thin sheet metal. There are two types:

- Thread forming, which are pointed and are used for soft materials.
- Thread cutting, which are blunt-ended and can be used on all materials.

■ TEST YOURSELF

Below are three products where similar or dissimilar materials have been joined together.

a) A desk light where the arm can be adjusted to different angles.
b) A flat-pack bookcase.
c) A climbing frame in a children's play area.

For each of the three products, give details of the methods of fixings that you would specify for use in each product. Give a reason for each choice (see also pages 34–35). **(6 marks)**

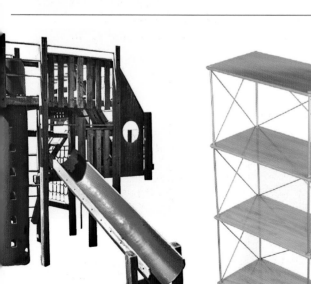

Fixing it all together [continued]

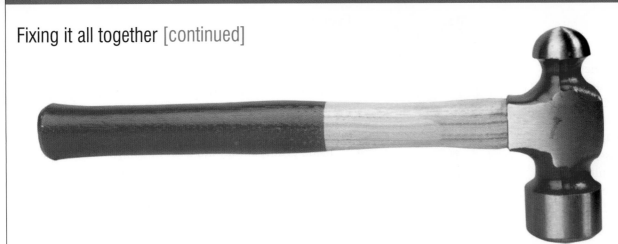

Nails and pins

Nails are metal pins with heads. They are used to join two pieces of wood and/or manufactured board together. The head stops the nail from being pulled through the top piece of wood or manufactured board if the two pieces are pulled apart.

There are many different types of nail that can be specified. These are the most common:

- Round wire nail: used for general carpentry.
- Lost head wire nail: used for general carpentry. The head is not exposed, so should be used where appearance matters.
- Oval wire nail: oval, so less likely to split wood.
- Panel pin: general purpose. The head can be punched beneath the surface of the wood and filled.
- Hardboard nail: used for hardboard. Made of copper, it will not discolour paint.

Rivets

Rivets are a quick and convenient method that is often used to join sheet metal together. They are usually made from aluminium, but also come in copper and stainless steel.

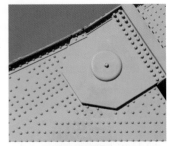

Common diameters are 3, 4 and 5mm and they are available in three different lengths. They are often used with backing washers. For blind riveting, access is only needed from one side.

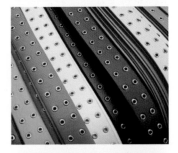

To fix rivets, a hole needs to be drilled that is just large enough to allow the rivet to pass snugly through it. The rivet is then inserted using a pop riveter or rivet pliers.

Fixings and fittings

There is a vast range of ready-made fixings and fittings available from DIY shops. Sometimes a manufacturer will design specific fittings for use with their own products. These fittings save time and money for manufacturers who do not have to invest in people and machinery to make traditional joints.

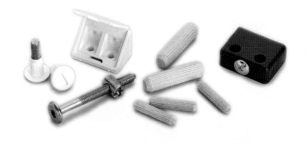

Knock-down fittings

Modern furniture is often made using manufactured boards and is available as self-assembly packs. This method of producing furniture has led to a revolution in KD or knock-down fittings. These are designed to be as 'user friendly' as possible.

KD fittings also mean that furniture can be produced in flat packs. In this way, storage space at shops and factories is saved, and transport costs from factory to showroom are reduced. Many items can be taken home from the shops by the customer, which saves on the cost of delivery.

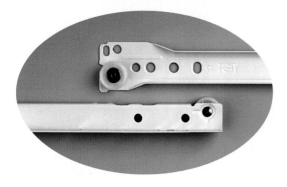

Fixings and fittings for wood

An example of KD fittings for wood are holes which are pre-drilled for screw fittings; plates and inserts already attached to the individual pieces; and multiple holes drilled for a choice of door and shelf positions. Many of these types of construction are held together using dowel pegs.

Fixings and fittings for metal

There is a wide choice of mechanical fastenings for metal and the method chosen will depend on a number of factors. The physical conditions in which the product will be used, for example the levels of humidity, and the physical characteristics needed by the joint, for example corrosion, are important.

Hinges

Hinges come in a variety of types for different uses. The main types are butt, piano, back flap, and flush. Rising butt and tee hinges are used on doors.

■ **TEST YOURSELF**

Modern flat-pack furniture that uses KD (knock-down) fittings offers advantages to both consumers and manufacturers of the furniture. Give *two* advantages for the:

a) consumer   (4 marks)

b) manufacturer.   (4 marks)

Fixing it all together [continued]

Fabric construction

The way in which fibres and fabrics are joined together can affect the appearance, texture and properties of the fabric.

Sewing

Sewing is the most common method of joining fabrics together. Hand stitching is used for tacking pieces close together and for attaching buttons. Sewing machines use two threads. Choosing the correct machine needle is important: it needs to match the thickness of the fabric.

Knitting

Knitting has a looped structure. Warp knits cannot be produced by hand knitting. They have yarns, which run 'up' the fabric, and they can be firm or slightly stretchy. In weft knitting, the yarns run across the fabric, forming loops with the row underneath.

Different knits

Different knitted structures produce fabrics with different textures, characteristics and appearances. The yarns used will also determine the final appearance of the fabric. For example:

- A simple single jersey structure is often used for T-shirts.
- A double jersey structure gives a thicker and less elastic fabric than single jersey.
- Rib-knit or ribbing gives a very elastic structure, useful for cuffs and garments that need to be close fitting.

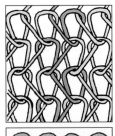

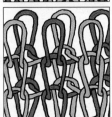

Top left:
Warp knitting.

Top right:
Single jersey.

Bottom left:
Double jersey.

Bottom right:
Rib-knit.

Knitted fabrics have a number of advantages over other methods of fabric construction. For example:

- Knits do not fray at the edge the way wovens do.
- Knits have much more stretch than woven, felted or bonded fabrics. This allows close-fitting garments to be designed that stretch with the body when it moves. They are therefore very comfortable to wear and used widely for leisure and sportswear.

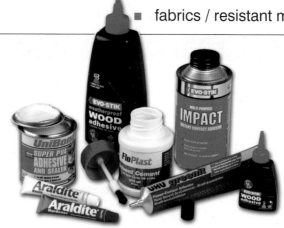

Adhesives

Adhesives hold two surfaces together in what is usually a permanent joint. As an increasing range of manufactured materials has been developed, so has the need for a range of specialised adhesives to join them. Designers need to make sure they specify the right one.

Bonded and felted fabrics

Bonded fabrics, sometimes called non-woven fabrics, are made from 'webs' or 'batts' of fibre held together by a resin adhesive, needle-punching or stitch bonding with thread. They are rarely used for the main body of a garment or other textile item but more often for interlinings and insulation, as well as disposable hygiene products, resin-free liners for floppy disks and artificial leathers.

Felted fabrics are an older type of non-woven structure and are made from a combination of moisture, heat, friction and pressure. Most carpet underlay is synthetic felt.

Bonded and felted fabrics can be produced more cheaply than woven and knitted fabrics but are not as strong.

Manufacturers use **interlinings** to strengthen and stiffen garment parts such as collars, cuffs and waistbands. When choosing an interlining, the designer must consider the weight and thickness of the fabric for which it is intended. A very thick and stiff interlining on a delicate fabric would not look nor feel right. Interlinings can be bought with an adhesive on one side that allows them to be ironed on, or they can be stitched on.

Applying adhesives

There are some important things to remember when applying adhesives. For example:

- Surfaces to be joined will need to be clean, dust free and dry.

- Equal or varying parts of resin, hardener or curing agent will need to be mixed together in the correct amount using a clean container or piece of card and a spatula.

- Even layers need to be applied to the surfaces to be joined.

- Excess adhesive needs to be wiped off immediately.

- The structure needs to be held securely while drying, using a clamp or masking tape if it is an awkward shape which needs to be supported.

Resistant materials

There is a wide variety of adhesives, both natural and synthetic. **Natural adhesives** (glues) are animal or vegetable-based products made, for example, from bones, fish and plant extracts. A glued joint relies on two things – a good fit and a clean surface. **Synthetic adhesives** are mainly polymers. They have strength and flexibility. Thermoplastic resin-based adhesives are softened by heat.

- **Epoxy resin** (Araldite) is very adaptable. It dries within 48 hours. It is best for joining metal to wood, metal to metal and acrylic to wood, metal and acrylic.

- **Urea formaldehyde** is a transparent thermosetting two-part adhesive.

- **Extramite** (formerly known as Cascamite) is a dry powder that is mixed with cold water.

- **PVA** (polyvinyl acetate) (Resin W) is a white, ready-mixed liquid. It is best for joining wood to wood and for joining expanded polystyrene to anything!

- **Contact adhesive** (Thixofix) is a synthetic rubber solution that sticks on contact. It is good for joining a wide range of materials: fabric to wood, metal, acrylic and melamine; melamine to melamine, wood, metal and acrylic; and rubber to rubber, wood, metal, acrylic, melamine and fabric.

- **Acrylic cement** (Tensol), used for plastics, has strong fumes.

- **Latex fabric** glue (Copydex) is a rubbery white solution that is best for joining fabric to fabric.

- **Hot glue** (Bostick) is an adhesive that provides an immediate bond as it cools.

■ **TEST YOURSELF**

Explain carefully the steps for using epoxy resin adhesive.

(5 marks)

Finishing off

Most materials need to have some form of finish applied to them. This adds to their properties by making them stronger, water-resistant, more durable, more attractive, etc.

Smoothing things over

Often, materials need to be prepared first so that they will take the finish well. The surfaces of resistant materials can then be stained, polished or painted, and given a plastic-based finish to add an attractive, shiny and tough outer layer of protection.

Files

Files are hand tools that are used to smooth metal, plastic and wood, and to enlarge holes. There are seven main shapes of file, and three main degrees of coarseness. As well as files, rasps and surforms can be useful.

Abrasive papers

The main types for wood are glass paper (buff), garnet paper (red) and aluminium oxide paper (green). Emery cloth and silicon carbide paper are used on metal. Silicon carbide paper is sometimes called 'wet and dry' paper. To be effective, abrasive papers need to be wrapped around a wooden block.

Fixing and filling

Small holes and gaps tend to occur where different materials are joined together. These need to be 'filled' in some way using body filler, plaster, etc.

Portable sanders can be used for rapidly producing a fine finish over a large surface area. After they have been sealed, surfaces can be finished using acrylic, cellulose or water-based paints.

Stains and varnishes

Stains can be applied to wood to change its colour, while still allowing the grain to show through. Oils and polishes can also be used on wood. Polyurethane varnishes provide an attractive and tough finish, and can be matt, satin or gloss.

Paint

Paint can provide a colourful and durable finish.

- Oil based paints are tough, but expensive. Polyurethane paints are particularly durable. Enamel paints are useful for detailed work, such as on small-scale product models.
- Water-based or emulsion paints are cheaper but less durable. Many are now waterproof.

Special graphic finishes for paper and card

In addition to the basic print processes, there are a range of special graphic effects that can be created. The four most common are:

- varnishing
- laminating
- embossing
- foil blocking.

Other special effects can include the use of metallic papers and inks, and holographic images. Specifying special printing effects can add considerably to the cost of the print run.

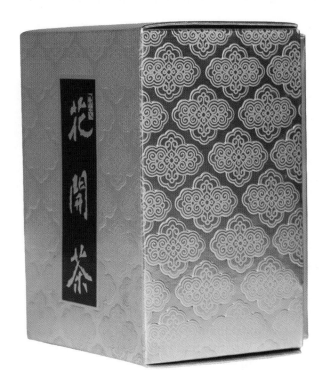

Varnishing

Varnishing is the addition of a thin glossy finish to protect the printed surface and make it look more attractive. There are four main types of varnish:

- Oil-based.
- Water-based.
- Ultra-violet.
- Spirit-based.

Spot varnishing is where only a particular area of the surface is varnished. This provides a contrast between the varnished and matt areas.

Laminating

Laminating is the addition of a thin plastic coating to achieve a high-gloss finish. Laminating is significantly more expensive than varnishing, but does provide better protection and quality of finish.

Encapsulation is a completely different process in which a document is placed inside a pre-gummed transparent plastic envelope and sealed inside.

Embossing

Embossing is where part of the surface of the paper or thin card is raised slightly to give a 3D effect. This can make a printed surface of a package or a book cover much more distinctive and interesting. Each sheet needs to be pressed over a steel die that carries the required shape.

■ TEST YOURSELF

1. Explain briefly why most materials need to have some form of finish applied to them. **(2 marks)**

2. Presentation models made from block modelling materials usually need to be filled. Explain what is meant by the term 'filled'. **(2 marks)**

3. After being filled, presentation models are usually finished. Give *two* examples of suitable finishes that might be applied. **(2 marks)**

4. Obtain *three* graphic products that have had a special finish applied. Suggest which methods have been used for each. **(6 marks)**

Finishing off [continued]

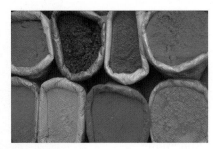

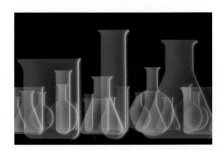

Dyeing

Dyeing is when all the fabric is completely immersed in dye solution. Dyeing is different to printing. In printing, colour is applied to the surface of the fabric only. Tie and dye and Batik are two traditional resist techniques for producing patterns with dye.

Dyeing can be done at one of four stages of the production process:

- Fibre.
- Yarn.
- Fabric.
- Completed garment or product.

Where do dyes come from?

Dyes can be made from **natural** or **chemical** sources.

Natural dyes are made from a variety of sources, such as plants and vegetables. They are very time consuming to make and, therefore, are rarely used today.

Natural dyes have a number of particular disadvantages in manufacture:

- It is very difficult to predict the final outcome of a natural dye.
- It is also difficult to mix the exact same colour again and again.
- Growing enough of the appropriate plant material would use up vast amounts of land.

Synthetic chemical dyes were developed in the 1850s. By the beginning of the twentieth century, synthetic dyes were used for most of the coloured textiles produced in Europe and North America.

Synthetic dyes have a number of particular advantages over natural dyes in manufacture:

- A wider range of colours can be created.
- Brighter colours are available.
- The dyes are cheaper to produce.
- Each colour is made to a scientific formula, so the same colour can be reproduced repeatedly.

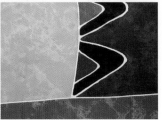

Fabric finishes

Finishing last

Finishes are applied to fabrics to make them more suitable for their intended use. Finishes need to be as inexpensive and durable as possible without compromising handling, strength or wearing properties. Finishing is usually the last stage of fabric processing.

There is a wide range of finishing processes that affect the performance of the fabric in different ways. Performance finishes can be divided into finishes that modify:

- the surface of the fabric
- the wearing properties
- aftercare characteristics
- appearance.

Modifying the surface

An example of a finish that modifies the surface of the fabric in some way is 'raising'. Fabrics can be brushed or raised by passing them over a large rotating brush. This raises the surface of the fabric. As well as making it feel softer, this traps more air so that clothes made from the fabric feel warmer.

Modifying wearing properties

Finishes can modify the wearing properties of a fabric. For example, fabrics can be made stain and soil resistant by applying a mixture of silicone and fluorine. This prevents grease and dirt entering the fabric, which is very useful for carpets and upholstery. A water-repellent finish can be applied to fashion rainwear to protect from light showers. However, a waterproof finish can completely prevent the penetration of water.

Modifying aftercare characteristics

Finishes can be used to modify aftercare characteristics. Fabrics such as cotton that crease badly can be treated using a process that involves combining the fibres with a resin. As well as making them look better when being worn, it makes them easier to iron after washing.

Treatments that prevent shrinkage also alter aftercare properties.

Modifying appearances

Finishes such as bleaching, smoothing and embossing are used to improve the appearance of fabrics.

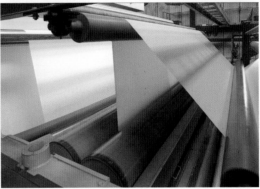

■ TEST YOURSELF

1. Discuss the advantages and disadvantages for a small-scale manufacturer of specialist, luxury textile products of using natural or synthetic dyes. **(8 marks)**

2. Below are photographs of three different uniforms made from polyester/cotton fabrics. What finishes might be applied to make them more suitable for their intended purpose?

   (3 marks)

Mechanical systems

Many products use mechanical systems and components to make them work.

Systems

Systems are groups of related components and events that work together to make something happen. In a system, inputs are transformed into outputs. The transformation process can be controlled through feedback about what the system is doing.

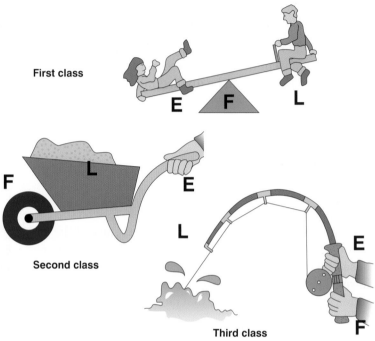

First class

Second class

Third class

Including mechanical components

Move it!

Mechanical systems help make machines more efficient by reducing the amount of energy required to use them. They can also change one type of motion into another.

Levers, cams, cranks and sliders use four different types of motion:

- Linear motion is movement in a straight line.

- Reciprocating motion is a repeated backwards and forwards movement in a straight line.

- Rotary motion is circular movement.

- Oscillating motion is a repeated left to right/right to left movement that follows a curved path.

Levers

There are three classes of lever: first; second; and third. The position of the load, effort and pivot or fulcrum decides which class a lever belongs to. All levers consist of a load, effort, pivot or fulcrum. These can be in different positions. Finding the correct position for the pivot is very important.

Cams

Cams are irregular-shaped wheels, or can be round with off-centre holes. They rotate around an axle and have a follower resting against them.

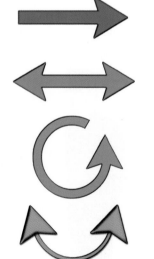

Cranks and sliders

Crank and slider mechanisms can be used to change reciprocating motion to rotary motion and to change rotary motion to reciprocating motion.

Linkages

When two or more levers are joined together, it is called a linkage. It is possible to connect two levers to move in opposite directions at the same time. Some pop-up mechanisms link different combinations of types of lever together.

Getting into gear

Gears are mechanisms based on the principle of the toothed wheel and axle. The driver drives the driven gear. When two gears are meshed, they turn in opposite directions.

- To turn them in the same direction, use an idler gear between them.
- A small gear meshed to a large gear decreases speed.
- A large gear meshed to a small gear increases speed.

When two or more gears are meshed together, it becomes a gear train. A compound gear train is made from pairs of meshed gears. They are for making large speed changes or getting different outputs moving at different speeds.

Top far left/middle:
Spur gears mesh together.

Top right:
Bevel gear.

Right:
Chain and socket on a bicycle.

Pulleys

Pulleys can make it easier to lift loads by altering the mechanical advantage. A pulley increases the amount of effort that can be put into lifting the load. Pulley systems can also move a load in a different direction to that in which you exert a force. There are a number of different types of pulley systems, for example:

- Single fixed pulley.
- Block and tackle.
- Belt drives.

■ TEST YOURSELF

Suggest what types of mechanical components might have been used in these products. . **(6 marks)**

Electronic systems

Electronic systems and components are used in a wide range of products.

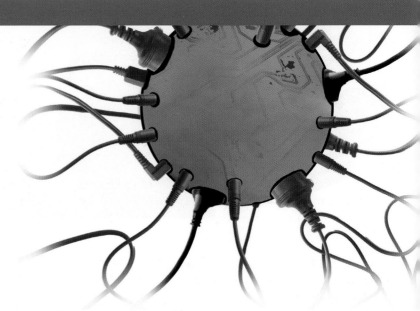

Electronic devices

Electronic devices are made up from a number of different components, each of which does a particular job. The components are arranged in a circuit. The circuit transforms a series of inputs into a series of outputs. Circuit diagrams are used to draw electronic circuits.

Incorporating electronic components

Switches

These turn the current on or off by making a break in the circuit. There are lots of different types of switches that can be used.

Resistors and capacitors

Resistors reduce the rate of flow of current around a circuit. Different resistors are used to produce different rates of flow.

Capacitors store a small amount of electrical charge that can be released at a particular time, a bit like a battery.

Resistors and capacitors are often used together. The resistor controls the discharge of the capacitor.

Transistors

Transistors can be used to amplify a current or switch it off.

LEDs

LEDs (Light Emitting Diodes) can be used to show that a circuit has current flowing around it, i.e. that it is on. They come in a range of colours and sizes, and can be made to flash on and off.

Resistor

Capacitator

Transistor

LED

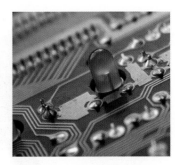

PCBs

Before a circuit is finalised, a prototype is usually made to test it out using 'breadboards'. When it has been tested, a **PCB** (Printed Circuit Board) is made. This involves etching the surface into a piece of copper and soldering the components on.

ICs

ICs (Integrated Circuits) combine a range of small electronic components that perform a specific purpose into one component. For example, a 555 timer IC is a timing device.

Meanwhile, **PIC** (Programmable Integrated Circuit) chips can be given instructions to perform different actions at different times.

Sensors

All circuits require **inputs**. Sensors are often used to trigger a circuit into action. There are two main types of sensors: **remote** sensors; and **physical** sensors.

Remote sensors react to change in conditions such as heat or light. Examples of these are light dependent resistors, temperature dependent resistors, infrared detectors, moisture detectors and magnetic field detectors.

Physical sensors require a force of some sort, and are essentially switches of one sort or another, either directly or by an action. There are lots of different types of switches, such as pressure, reed and mercury switches.

DC motors

All circuits also require **outputs**. Motors convert electrical energy into a rotational movement. They come in a wide variety of sizes, and their speed and direction can be controlled.

Sound and light

Buzzers and loudspeakers can be used to convert electrical pulses into sound waves. As well as LEDs, small DC light bulbs can be lit.

Electromagnetic devices

Solenoids produce small, simple linear movement. Magnetism is used to repel a piece of metal. This can be linked to a more complex mechanical system.

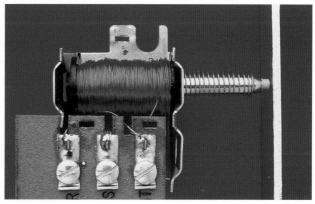

■ **TEST YOURSELF**

Study the electronic products on the right. Suggest *three* types of electronic components that might be found in each. **(9 marks)**

Computer trackball mouse.

Heart-rate monitor.

PCB	IC	Sensor

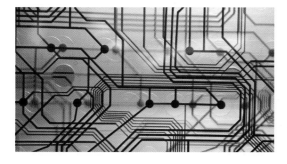

Materials and components

Safety matters

Workshops and studios can be dangerous places.
There are lots of tools and equipment that can cause a great deal
of harm if safety precautions are not followed.

It's important to think about the tools and equipment you use
and the space you use them in.

Using hand tools

Even if you are only using hand tools, there are many possible hazards to be
aware of. Saws and chisels have very sharp edges that can easily cut and
skin is less resistant than wood, metal or plastic! Adhesives, spray paints,
craft-knives and even scissors can cause injury if not used with caution.

Using power tools

A designer's workshop will probably
include a variety of electrically-
powered drills, small lathes and
hand-held tools. All these are
potentially dangerous if mishandled.
Always stop to think about the
specific safety warnings your teacher
has given you for each one. Also
remember the general workshop
rules that need to be observed.

Using ICT

Remember too that there are safety
considerations involved when
using ICT. Never open the computer
casing. The components inside carry
a high voltage charge. If you use
computers for long periods, you may
start to suffer from backache, eye
strain or aching arms.

Placing tools and equipment

Careful planning of the layout of
workshops and studios can help
reduce the risks of injury. The
positioning of some tools and
equipment close together can be
dangerous.

- Placing a computer next to a sink
  would not be a wise idea, as
  splashing water might lead to
  electrocution.

- The placing of power points and
  trailing cables also needs close
  attention.
- Storage systems need careful
  thought. Is it easy to reach
  dangerous items? Are any large
  piles of material likely to fall
  over?

Maintaining air quality

Facilities for ventilation and fume
and dust extraction are also essential
considerations.

- Dangerous fumes can be given
  off when plastic-based products
  are cut, or when two chemical
  components are mixed.
- Adhesives, spray paints and
  varnishes can be dangerous if
  inhaled.
- Fine dust from cut materials can
  also be a health problem if not
  extracted.

Looking after yourself

Protective clothing, such as goggles, ear protectors and overalls, needs to be readily available. There also needs to be facilities for regular washing.

Food-wise

If you are making a food product, make sure you follow the hygiene rules set for the food preparation you are working in.

In particular, make sure you avoid any food contamination through contact with other foods or work surfaces that have not been thoroughly cleaned.

Make sure you wash your hands thoroughly and regularly to help avoid bacterial infection.

■ TEST YOURSELF

1. On the right is a photograph of a desk-top drill and a craft-knife below. Identify *two* important safety requirements to remember when using each. **(4 marks)**

2. Briefly describe the potential health risks involved in working at a computer screen for long periods of time. Give *three* possible symptoms and the most likely cause of each. **(6 marks)**

3. Identify *two* important safety considerations when planning the layout of a small workshop. **(4 marks)**

Hazards and risks

Accidents in manufacturing can have serious consequences. Unsafe acts or conditions are called hazards.

There is an important difference between a risk and a hazard. A risk tells you how likely it is that a problem will happen. A hazard is the problem itself. If something is risky, it means there's very likely to be a hazard involved.

Assessing the risk

Risk assessment helps to make manufacturing safe by minimising the risk of hazards. Risk assessment is about identifying what hazards might happen, how likely they are to happen, and what the result would be if they did happen. It works out where the risks of hazards are in a process, and how to make things as safe as possible. Different processes and areas of the workplace may be identified as being high, medium or low risk.

For example, a risk assessment for the process of cutting out fabric with a band knife might look like this:

What are the risks?
1 The operator may cut their hands with the knife.
2 Dust from the fabric may get in their eyes, nose or throat.
3 The operator may be distracted by someone else and this may cause an accident.

How can these risks be reduced?
1 The operator must wear chain mail gloves.
2 The operator must wear a mask and goggles.
3 Paint lines on the floor to identify the cutting workspace and display clear warning signs that prohibit others from entering that area.

Employer responsibility

There are a wide range of regulations and guidelines that manufacturers must follow when asking employees to work with various tools, machines and materials, or when undertaking various processes.

Employers must justify the level of precautions adopted to a Health and Safety Inspector. Once the risks have been identified they must be controlled and monitored.

Employee responsibility

Employees must abide by all agreed health and safety regulations.

Choosing and starting projects

Identifying suitable design and make projects for yourself is not easy. A carefully chosen project is much more likely to be interesting and easier to complete successfully. Investing time and effort in choosing a good project makes progress a lot easier later on.

Project contexts

the high street

transport

communications

clothing

leisure

security

food

health

education

the home

energy

the natural environment

Making a start

There are a number of different ways in which you might start a project. Your teacher may have:

- told you exactly what you are required to design for a specific situation and market
- stated the product you must design, but not defined the target market and context
- given you a range of possible design tasks for you to choose one from
- left it up to you to suggest a possible project.

If you have been given a specific task with a defined situation and market, go straight to page 50.

Observe people carrying out everyday tasks. What do they find difficult or unpleasant? Are there potential opportunities for new or improved products that would help make life easier or more enjoyable for them?

❝ *don't forget*

Explain why you think this is a good opportunity for design. What are the potential users' important needs and wants? What problems will you need to solve?

Remember, it's important that what you design is suitable for production, even if only in small numbers. It can't be just a one-off item. You will need to show some plans for your product to be factory made.

Defining the target market

The target market is the group or type of people your product is intended for (see also page 88). Who are the potential users? How and when will the product be used? If you don't know the answers to these questions, how can you find out?

Create an initial user profile for your product. Find pictures that show the age, interests, buying habits and lifestyles of your target market.

Stating the outcomes

This does not mean what the final design would be, but the form your final prototype might take, e.g. a working model or product, a garment, together with appropriate presentation and manufacturing drawings, etc. Think carefully about the following:

- Does the success of the project depend on important information you might not be able to get in the time available?
- Might it be expensive or difficult to make?
- Do you have access to the tools and materials likely to be required?
- Will you be able to work out how it could be manufactured in quantity?
- Are there good opportunities for you to use ICT, including CAD-CAM?
- Apart from the main product, will you also be designing accessories, instruction manuals or promotional materials?

Make sure you discuss your project ideas with your teacher to check you are going in the right direction.

A very important consideration is the testing of a prototype of some sort, and of your final design. How would you be able to do this? Could ICT be used?

■ Identifying your own problem

If it's up to you to choose your own starting point for a design problem to solve, you will need to start by making a list of:

- potential local situations/environments you could visit where you could do some research into the sort of things people there might need (e.g. a local playgroup, a small business, a hospital or sports centre, etc.)

- people you know outside school who might be able to help by providing information, access and/or advice.

Arrange to visit some of the situations you've listed. Choose the ones that you would be interested in finding out more about. Make contact with the people you know, and get them interested in helping you. Tell them about your D&T course and your project.

For each possible situation you should:

- visit the situation or environment

- make initial contact with those whose help you will need

- check whether the success of the project depends on important information you might not be able to get in the time available.

With a bit of luck, after you've done the above, you should have a number of ideas for possible projects. Aim to choose a problem that:

- is for a nearby situation you can easily use for research and testing

- you can get some good expert advice about

- you feel interested and enthusiastic about!

in my design folder

✓ My project is to design a...
✓ I am particularly interested in...
✓ I have made a very good contact with...
✓ My prototype can be tested by...
✓ My final outcome will include...
✓ I could use ICT to...

Where do good ideas come from?

Having brilliant new ideas isn't easy, but it is something that everyone can do. All you need are a few simple techniques – and a lot of practice. The challenge is to make someone say, 'Wow!' when they look at what you've done!

The zip fastener dates back to an invention by a Swede, Gideon Sunback, in 1913. Zip fasteners were first used for pilots in the First World War, and became increasingly used in everyday clothing during the 1920s.

Creativity and innovation: finding the 'Wow!' factor

Creativity involves designing and making something that didn't exist before. At one level it might only be slightly different to an existing product or system – a change of shape or colour perhaps, or it might be a completely new way of doing something – a product that no one else has come up with before.

It's one thing to have an idea for a brand new type of product, but another to get it made and bought in quantity, i.e. to become successful. New ideas that become established in the market are known as **innovations**.

Sir Clive Sinclair successfully produced and marketed the first slimline 'pocket' calculator in 1972: previous hand-held calculators had been bulky and heavy.

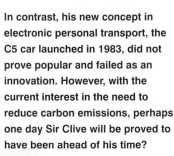

In contrast, his new concept in electronic personal transport, the C5 car launched in 1983, did not prove popular and failed as an innovation. However, with the current interest in the need to reduce carbon emissions, perhaps one day Sir Clive will be proved to have been ahead of his time?

Think different!

Before they become accepted, new ideas often seem most unlikely – wacky, crazy and eccentric. To design and make something that didn't exist before demands courage, and a high degree of risk. There's no certainty it will work. Successful designers are those who seem to have fun 'playing' with ideas. If your designs surprise and delight someone and make them say 'Wow!', then the chances are you're on to something creative and potentially innovative. So don't be afraid to **think different**!

In your coursework projects you are not expected to come up with something innovative (though it would be great if you did!) However, you do need to design and make something that is imaginative and different in some way.

Also, remember that you can (and should) aim to be creative in all aspects of your coursework. So for example, you might:

- **investigate** unusual sources of information that produce unexpected results
- **experiment** with new materials and processes
- **present** your work in a different way.

The paper clip as a means of holding sheets of paper together without tearing them was patented in 1900 by Johann Vaaler, a German.

The Post-it note, at the bottom of page 50, was invented in 1974 by Arthur Fry who worked for 3M, an adhesives manufacturer. It was originally intended as a page marking device.

First thoughts: getting going

When you start designing you need lots of ideas – as many as possible, however crazy they might seem. But the worst thing is to sit staring at a blank piece of paper or computer screen. Below are some simple methods for getting going.

Phone a friend!

If you feel yourself getting stuck, then don't keep it all to yourself. Talk about it to the person sitting next to you, phone someone you know to discuss the idea, or e-mail a friend. Often simply trying to explain something to someone else helps to clarify the problem in your mind.

❝ *don't forget*

As usual, it is essential to record all your ideas and thoughts.

Much of your work, particularly early on, will be in the form of notes. These need to be neat enough for someone else to read.

If this is your final examination coursework project, check with your teacher to ensure that what you are proposing to design and make will fulfil the assessment criteria.

There's nothing new under the sun!

There is a saying, 'There's nothing new under the sun'. What this means is that most things have been invented already, so there's not much point trying to invent something completely original. The vast majority of good ideas develop by looking at things that exist already and changing and combining them in a way that's never been done before.

Brainstorming

Brainstorming (or 'idea-storming' as it is now sometimes called) is a popular technique. A session needs to be done as a group activity and no criticism of ideas is allowed! Essentially, brainstorming widens your thinking and gets you out of the habit of dismissing crazy ideas too early on.

Mind mapping

Mind mapping can help you organise the different parts of a problem. It is a good follow-on from brainstorming. Start with a central idea and work out how its parts radiate from the centre. Use different colours and styles of lines and arrows to plot the relationships between the parts. You may need to do several versions to unravel the diagram, in order to reveal a clear structure in which as few lines as possible cross each other.

Take a break

Strange as it may sound, 'giving up' can be one of the most successful techniques for generating good ideas! Providing you've spent some time wrestling with the problem and some possible solutions, go away and do something else. Without you realising it, your brain will keep on working and, surprisingly often, a great idea will suddenly just 'pop' into your mind. For this reason, it's always a good idea to keep a small pocket notepad with you at all times so you can jot and sketch ideas down.

It probably never happened, but there's a well-known story that the Greek scientist Archimedes was sitting in the bath when he suddenly conceived the principle of the water pump. It is said that he exclaimed, 'Eureka! I've got it!' hence the notion of having a 'Eureka moment' when everything suddenly becomes clear.

What do I need to find out more about?

You will need to find out as much as you can about the people and the situation you are designing for. To do this, you will need to identify a number of different sources of information to use for your research. Studying similar, existing solutions can be a good place to start.

Starting questions

Begin by writing down what you already know, or think you know, about the design opportunity. Then make a list of questions you will have to find answers to. You should find the following prompts useful:

- Why…?
- When…?
- Where…?
- What…?
- How many…?
- How often…?
- How much…?

- Why is this product needed?
- How many people will use it?
- When and where is it used?
- What special features does it need?
- What materials would be most appropriate?
- Where will it be kept when not in use?

Sources of information

Next, carefully consider and write down the potential sources of information you might be able to use in order to discover the answers to your starting questions. Refer to the research methods on the next page. Be sure to give specific sources as far as possible (i.e. name names).

Across your research you will need to aim to obtain overall a mixture of:

- **factual** information, e.g. size, shape, weight, cost, speed, etc.
- information which will be a matter of **opinion**, i.e. what people think and feel about things, their likes and dislikes, what they find important, pleasing, frustrating, etc.

❝❝ *don't forget*

Write down what you need to find out more about and how you could obtain the information.

You need to identify a number of sources of information (e.g. user research, existing solutions, expert opinion, information search).

Remember to record what you discover. Make sure your work is clearly and attractively presented.

Of all the research methods, user-research tends to be the most effective and useful, so you are highly recommended to include some in your investigation. Some form of personal contact is essential to a really successful project.

It isn't necessary to use all the research methods in any one project, but you certainly must show that you have tried a selection of them.

■ Research methods

Product analysis and evaluation

Analyse and evaluate existing solutions to the problem, or similar problems. What works well and what could be improved? What could be completely changed?

Site study

Document the environment in which the product would be used. Which of the following will be relevant?

- Historical and geographical factors.
- Sociological, economic, political information.
- Layout, facilities.
- Sizes and spaces.
- Atmosphere – light, colour, texture.
- Location and the surrounding environment.

User research

Observe and consult people who are directly involved in the problem. You need to find out about:

- the things they do, and
- the way in which they do them.

You could also undertake a small survey.

User trips and trials

Record your own impressions of the problem. Are there any relevant activities you could try out for yourself to gain first-hand experience? Do you have any recollections of any previous similar experiences you have had?

Expert opinion

Contact people you know who could give you expert professional advice on any aspects of the problem. If you don't know immediately of anyone, how might you set about finding somebody?

Information search

What information about the problem, or a similar problem, is already available in books, magazines, on TV programmes or the internet?

If you don't already know that such information exists, where could you go to look for it?

Think carefully about the best ways to record the information you discover.

In conclusion

When most of your investigation work has been completed, you will need to draw a series of conclusions from what you have discovered. What have you learned about the following things:

- The sort of people who are likely to be using the product.
- Where and when they will be using it.
- What particular features it will need to have.

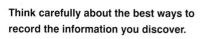

in my design folder

✓ The key things I need to find out about are...
✓ The research methods I am going to use are...
✓ I will be talking to the following people about my project...
✓ I will use ICT to...

in my design folder

✓ From my research I found out...
✓ I have discovered that...
✓ My conclusions are...
✓ I have kept my research relevant by...
✓ I found ICT helpful when...

Specify

An initial design criterion is a series of statements that outlines the possibilities and restrictions of the product at the start. This is likely to change and become more specific as more is understood about the problem and possible solutions.

A final product or manufacturing specification includes exact details about the features and appearance of the final design, together with its materials, components and manufacturing processes.

Writing a design specification

After you have done some research and had some first thoughts, you will be able to develop your **initial design criteria** into a more detailed **design specification**. This defines the things about the product that are fixed and the things that you are free to develop and change.

The conclusions from your research and early ideas should form the basis of your design specification. For example, if in the conclusions to your investigation you wrote:

'From the measurements I made of a number of people's arm lengths, I discovered that the best size for an arm-rest would be between 250mm and 400mm long.'

In the specification you would simply write:

'The arm-rest should be between 250mm and 400mm in length.'

Fixing it

Some statements in the specification will be very specific, e.g.: 'The toy must be red.'

Other statements may be very open ended, e.g.: 'The toy can be any shape or size.'

Most will come somewhere in between, e.g.: 'The toy should be based on a vehicle of some sort and be mechanically or electronically powered.'

In this way, the statements make it clear what is already fixed (e.g. the colour) and what development is required through experimentation, testing, and modification (e.g. shape, size, vehicle-type and method of propulsion).

The contents of the specification will vary according to the particular product you are designing. On the next page is a checklist of aspects to consider.

❝❝ *don't forget*

You might find it helpful to start to rough out the design specification first, and then tackle the conclusions to your research. Working backwards, a sentence in your conclusion might need to read: 'From my survey, I discovered that young children are particularly attracted by bright primary colours.'

It's a good idea to use a word processor to write the specification. After you've written the design specification new information may come to light. If it will improve the final product, you can always change any of the statements.

Make sure you include as much numerical data as possible in your design specification. Try to provide data for anything that can be measured, such as size, weight, quantity, time and temperature.

■ Specification checklist

Use and performance

State the main purpose of the product – what it is intended to do. Also define any other particular requirements, such as speed of operation, special features, accessories, etc. Ergonomic information is important here.

Size and weight

The minimum and maximum size and weight will be influenced by things such as the components needed, and the situation the product will be used and kept in.

Appearance

What shapes, colours and textures will be most suitable for the type of person who is likely to use the product? Remember that different people like different things.

Safety

A product needs to conform to all the relevant safety standards. Which of them will apply to your design?

- How might the product be misused in a potentially dangerous way?
- What warning instructions and labels need to be provided?

Manufacturing cost

The specification needs to include details of the total number of units likely to be made and the size and variations of batches.

Maintenance

State how frequently different parts of the product need to be maintained.

Life expectancy

Define how long the product should remain in working order, providing it is used with reasonable care.

Environmental requirements

To ensure your product will be made in the most environmentally friendly way, you might decide to:

- not specify a particular material because it can't be easily recycled or reused
- state the use of a specific manufacturing process because it consumes less energy.

Other areas

Other statements you might need to make might cover special requirements such as transportation and packaging.

Writing a final product or manufacturing specification

After you have fully developed and completed your Product Design, you will need to write a final detailed product specification. It should include precise statements about the materials, components and manufacturing processes needed to ensure that the manufacturer is able to make a repeatable, consistent product.

Your final product will need to be evaluated against your design specification to see how closely you have been able to meet its requirements, and against your product specification to see if you have made it correctly.

in my design folder

✓ My design will need to...
✓ The requirements of the people who will use it are...
✓ It will also need to do the following...
✓ It will be no larger than...
✓ It will be no smaller than...
✓ Its maximum weight can be...

✓ It should not be lighter than...
✓ The shapes, colours and textures should...
✓ The design will need to conform to the following safety requirements...
✓ The number to be printed or made is...
✓ The following parts of the product should be easily replaceable...
✓ To reduce wastage and pollution it will be necessary to ensure that...

Inspired by design

Designers are often inspired by everyday objects, products from other times and cultures, nature and geometry. They frequently take shapes, forms, colours, textures, structures, mechanisms from something that exists already and apply them to the product they are developing.

To inspire the user, designers also consider the emotional impact a product will have.

What's it like?

Studying the way things work will give you lots of starting points for your own designs. A mechanism, hand-grip or pattern design used on one product, with a little modification, could be just the perfect solution for a very different product.

Study the photographs on this page, the other products illustrated throughout this book, and in the world around you! Find some inspiration for your ideas in other things that:

- open and shut
- expand and collapse
- inflate and deflate
- twist and turn
- fold and unfold
- lock and unlock
- spring up and jump
- contain or extend
- make interesting sounds
- feel nice or nasty.

What does it feel like?

To stimulate new insights, imagine you are one of the products you have been studying. What does it feel like? What makes it happy or sad, excited or bored, calm or restless?

What does it remind you of?

Do any of the products remind you of places you've been to, people you've known, or things you've read about or seen? What words, sayings, songs, film or book titles make reference to the product?

Other times and cultures

A good source of ideas can be to look at artefacts and products from other times and cultures – where they often do similar things in different ways. How could such solutions be adapted?

Back to nature

Natural shapes and forms are often a great inspiration for a designer: the structures of birds, animals, vegetation and flowers, and the way they look and behave, can provide much stimulation and ideas for solutions.

Ideas that really count

Rectangles, curves and other shapes and solids can be combined together in arrangements and proportions that are pleasing to the eye. Mathematical sequences also form the basis of designs, particularly in terms of developing pleasing proportions of shape and form.

The Golden Section

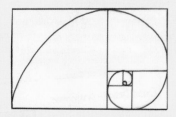

The ancient Greeks developed a mathematical solution for creating ideal proportions called the 'Golden Section', based upon a rectangle with a proportion of 1 to 1.618, or 1:1.6. They used it to work out the height and width of their buildings.

Leonard Fibonacci (1170 –1250), an Italian mathematician, devised a formula known as the **Fibonacci series**: 0, 1, 1, 2, 3, 5, 8, 13, 21, 34, 55, 89, 144, etc. Each number work is the sum of the two previous numbers. Choose any number in the series, divide it by the previous smaller number and the answer will be approximately 1.618. This sequence is often found in nature, for example in the stems of flowers and branches of trees.

Getting emotional

People don't just use products, they react to them **emotionally**. How will your design make them feel? Annoyed, irritated and frustrated, or delighted, pleased and satisfied?

As well as being easy to use, will its shapes, forms, colours, textures, sounds and smells make it pleasurable to use? Will it make the user feel more glamourous, attractive, popular, adventurous, intelligent, sophisticated, young, cool, up-to-date, in control, etc?

How well does your design achieve beauty, pleasure and simplicity of use? Will it make people smile, feel good, and say, 'Wow! I want one'?

How do you think each of these watches makes the wearer feel?

A good way to explore the possible look and feel of a new product is to create mood, image and material sample boards. These can be used for personal reference or to discuss with potential users to get their reactions.

Mood boards are a collage of objects, colours, textures and words that combine together to create a particular feeling.

Material sample boards combine samples of real materials together to explore combinations of surface textures and patterns.

Image boards bring together the things certain groups of people (i.e. your target market) have, do and aspire to in terms of their lifestyle.

Transforming ideas

When you've got some basic ideas sorted out, you need to develop them quickly in more detail. To help you do this, use a wide range of drawing techniques such as plans, elevations and 3D sketches, as well as words and numbers. This will help you develop and transform your ideas into exciting and imaginative products.

Second thoughts: getting down to details

When asked to design something, you tend to start with a hazy image in your mind of what it might be like. You grab a pencil and piece of paper to try and sketch out what you've seen in your mind's eye. But often what appears in front of you isn't at all like what you were thinking of! Being able to sketch effectively can be learned, just like you learned to write words.

Working on paper, begin to develop your ideas in more detail. Remember to use a range of drawing techniques such as plans, elevations, 3D sketches, as well as words and numbers to help you show your ideas. These development sheets will also help you to explain and discuss your ideas with other people.

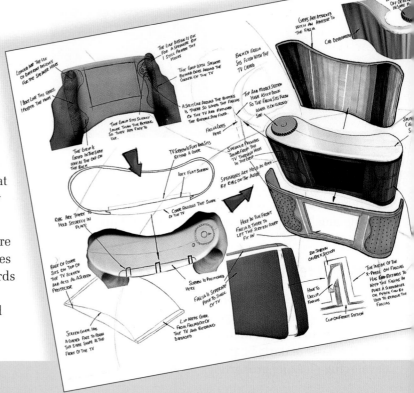

◼ Communicating your ideas

Communicating your ideas clearly and effectively through labelled drawings will help you to:

- visualise the ideas that you have in your head
- record your ideas and your reasons for developing the product the way you have
- explain your ideas to others, including your teacher and the examiner.

Sketching

The drawing technique you use needs to be quick and clear. Sketches should be freehand – rulers should not be used as they take time and can restrict your design work to straight-line shapes. Draw in 3D or use plans and elevations. Use colour and shade only if it helps to explain your ideas, not just to decorate the drawing. Use written notes to help explain and comment on your ideas.

Orthographics

Plans and elevations drawn together are known as orthographic projection. Make sure you follow the conventions for dimensioning. Sections through the product can often help to explain constructional details, as can exploded drawings.

System diagrams

Use system and circuit diagrams to show how different parts of your product relate to each other.

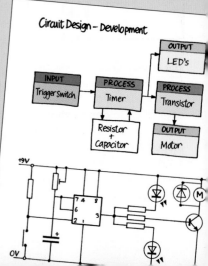

Transformations

Most design development involves some form of transformation – gradually changing things until they become something else. Try some of these transformation techniques to find ways of making your initial ideas more interesting, unusual and exciting.

What if, and why not...?

This is another technique to get you exploring new and different directions. Look at what you've got so far, and ask yourself questions such as:

- what if it were rounder, taller, a different colour, etc.
- what if it worked the other way round, upside down, inside out
- why not use fewer parts, more sustainable materials, etc?

Make sure your ideas still meet the requirements of your design criteria or specification.

2D transformations.

3D transformations.

Developing the lamp base and shade

Fatter · Thinner/Taller · Rounder · Curvier · Conical · fabric · plastic · paper · ceramic · metal

Sideways · Triangular · glass · wood · Low energy? · Tungsten?

in my design folder

✓ I chose this idea because...
✓ I developed this aspect of my design by transforming...

Take a chance

Can't decide which idea is the right one to take forward? Then don't spend forever worrying about it: just choose one and go for it. Your intuition, or 'gut-feeling' as it's sometimes called, will see you through. Successful designing is about making the best of a good idea. And if you still really can't make the decision, throw a dice and let fate make it for you!

Less is more

Many design ideas suffer from being over complicated: too many features, components, complex shapes that somehow just don't seem to work together. The best designs are often the result of '**letting go**' of what in themselves may be good ideas to produce a simpler, easier-to-use, more elegant solution. This isn't easy as our ideas become precious to us, and we are reluctant to throw them away.

From soft modelling to working mock-ups

'Soft' models are quick and cheap to make and can be easily changed as you develop your ideas. Working prototypes are needed to test out specific features of your design.

Material- card   Scale- 1:10
Purpose - to rapidly explore initial ideas in 3D

Materials - wire and fabric
Scale - 1:10
Purpose - to explore different shapes which can be made in tubular steel

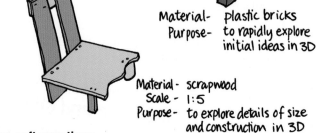

Material- plastic bricks
Purpose- to rapidly explore initial ideas in 3D

Material - scrapwood
Scale - 1:5
Purpose - to explore details of size and construction in 3D

Soft modelling

At some stage you need to start to try and get a better feel of how your design might look and work. To do this, you might find it useful to make a series of **soft models**.

Just as the idea for your next sketch usually occurs to you while you are working on your current drawing, so the idea for your next soft model will probably come to you as you make each one. Or maybe it's a re-working of your current construction. That's the great thing about soft models: they are **quick to make** and can be **easily changed**, modified or adapted as new possibilities emerge.

Going soft

Soft models are made from cheap and simple materials such as paper, card, foil, pipe-cleaners, balsa wood, straws, plasticine and pieces of fabric, fixed together using masking tape, paper clips, glue-guns, thread, string and rubber bands.

Your soft models will not usually be made in the materials the final product might be made in. They might also be at a different scale – smaller or bigger. Some might not even be of the whole idea, but just a part of it.

Remember to take a sequence of photographs of your models as they develop. Add these to your folder to help tell the story of how you developed your design idea.

Some soft questions

While you are making your soft models, the sorts of questions that you might be asking yourself are:

- is the overall size about right
- do the shapes, forms and proportions look good
- are the structures strong enough
- how well does it all fit together
- whereabouts would any displays and controls be placed
- what would it be like to use
- what materials might the final product be made out of
- am I still happy with the idea
- what problems will need to be sorted out to make it work properly
- what might be completely different about it?

While you are making your soft models, you should also continue to sketch and make notes as your ideas develop.

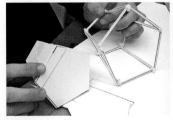

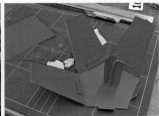

Making working mock-ups

Eventually, you will be in a position to bring your ideas into sharper focus by making some form of **working mock-up**. Think carefully about exactly what aspect of your idea you want to test out and about the sort of model that will be most suitable.

Some of your tests might involve obtaining other people's **opinions**. Try to also devise some objective tests to carry out on your mock-up involving **measuring** something.

Like your soft models, your mock-ups may need to be made using different materials and at a different scale to the final product.

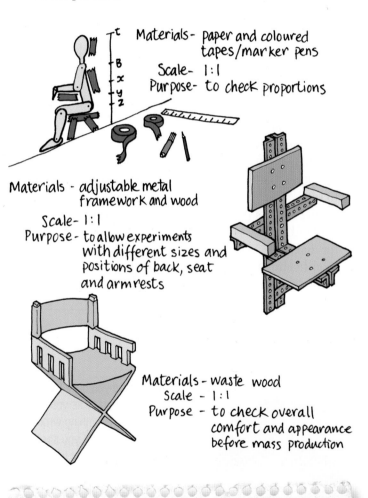

Materials – paper and coloured tapes/marker pens
Scale – 1:1
Purpose – to check proportions

Materials – adjustable metal framework and wood
Scale – 1:1
Purpose – to allow experiments with different sizes and positions of back, seat and armrests

Materials – waste wood
Scale – 1:1
Purpose – to check overall comfort and appearance before mass production

in my design folder

✓ To evaluate my ideas I decided to make a mock-up which...
✓ The way I tested my mock-up was to...
✓ What I discovered was...
✓ As a result, I decided to change...
✓ I used ICT to...

Working to scale

One of the first decisions is what size to make the soft model or working mock-up. Here are some commonly used scales:

- A small intricate item – 2:1 (i.e. twice as big).
- A hand-held object – 1:1 (i.e. full size).
- A piece of furniture – 1:10.
- An exhibition stand – 1:50.

At a scale of 1:10, 1 unit represents 10 units in reality. A model that is 100mm tall would show an object which is actually 1,000mm high.

It is a good idea to check the sizes and thicknesses of materials you have available. These might influence the exact scale you end up choosing.

Recording the results

Make sure you make some notes about the circumstances in which the tests were undertaken, and record your results. Write down some clear statements about:

- what you wanted the prototype to test
- the experimental conditions
- what you discovered
- what decisions you took about your design as a result.

Many mock-ups

Following your first mock-up you may decide to modify it in some way and test it again, or maybe make a second, improved version from scratch. Make sure you keep all the mock-ups you make, and use a digital camera to take photographs of them being tried and tested.

Sometimes you will need to go back to review the decisions you made earlier and, on other occasions, you may need to jump ahead for a while to explore new directions or to focus on a particular detail.

Bringing ideas together and resolving conflicts

Always keep the 'big picture' in mind, i.e. the whole of your product. Solving one part of the problem may have an impact on a different part of the design, and compromises are likely to be needed if everything is going to work well together.

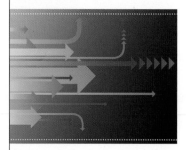

What's the plan?

The final prototype is very important. It presents your proposed design solution, rather than the process you used to develop it. Careful planning is essential. You will also need to be able to explain how your product could be manufactured in quantity.

Going with the flow

The final making is not just about the product you construct at the end of the project – it includes things like test pieces, samples, soft models, mock-ups, moulds, templates and jigs. Computer-aided manufacturing counts too!

Some planning is needed to ensure everything that needs to be finished is completed successfully and on time.

The main things you need to consider before you start making are:

- what materials will you need, and are they in stock
- what tools and equipment will you need, and are they available and working? Do you need to make any templates or jigs first?
- are there special skills involved you will need to practise
- how will you ensure your work is of sufficient quality
- how long have you got to do the work – hours, weeks or months?

You will need to provide evidence of this planning, such as flow charts, cutting lists, patterns, drawings, etc.

Quality counts

As your making proceeds, you will need to check frequently that your work is of an acceptable quality. How accurately are you working? How can you judge the quality of the finish?

If you are making a number of identical items, you should try and work out ways of checking the quality through a quality control sampling process (see page 122).

Keeping a record

Write up a diary record of the progress you made while making. Include references to:

- things you did to ensure safety
- the appropriate use of materials
- minimising wastage
- choosing tools
- practising making first
- checking that what you are making is accurate enough to work
- asking experts (including teachers) for advice explaining why you had to change your original plan for making
- modifications you made as you worked, and why you made them.

" *don't forget*

You may find you have to change your plans as you go along. There is nothing wrong with doing this, but you should explain why you have had to adjust your schedule, and show that you have considered the likely effect of the later stages of production.

What needs to be done by:
- next month
- next week
- next lesson
- the end of this lesson?

How many?

What you have designed should be suitable for manufacture. You should discuss with your teacher how many items you should attempt to make. This is likely to depend on the complexity of your design and the materials and facilities available to you. It may be that you only make one item, but also provide a clear account of how a quantity of them could be manufactured.

Making it better

Try asking the following questions about the way your design might be manufactured in quantity:

- What work operation is being carried out, and why? What alternatives might there be?
- Where is the operation done, and why? Where else might it be carried out?
- When is it done, and why? When else might it be undertaken?
- Who carries it out, and why? Who else might do it?
- How is it undertaken, and why? How else might it be done?

> Remember that manufacturing is not just about making things. It is also about making products:
>
> - simpler
> - quicker
> - cheaper
> - more efficient
> - less damaging to the environment.

Don't forget that there is also a high proportion of marks for demonstrating skill and accuracy, overcoming difficulties and working safely during the making.

■ Making it in quantity

Explain how your product would be manufactured in quantity. Work through the following steps:

1. Determine which type of production will be most suitable, depending upon the number to be made.

2. Break up the production process into its major parts and identify the various sub-assemblies.

3. Consider where jigs, templates and moulding processes could be used. Where could 2D or 3D CAM be used effectively?

4. Make a list of the total number of components and volume of raw material needed for the production run.

5. Identify which parts will be made by the company and which will need to be bought in ready-made from outside suppliers.

6. Draw up a production schedule that details the manufacturing process to ensure that the materials and components will be available exactly where and when needed. How should the workforce and work space be arranged?

7. Decide how quality control systems will be applied to produce the degree of accuracy required.

8. Determine health and safety issues and plan to minimise risks.

9. Calculate the manufacturing cost of the product.

10. Review the design of the product and manufacturing process to see if costs can be reduced.

 More information on all these topics can be found on pages 120–123.

in my design folder

✓ I planned the following sequence of making...
✓ I had to change my plan to account for...
✓ I used the following equipment and processes...
✓ I paid particular attention to safety by...
✓ I monitored the quality of my product by...
✓ My product would be manufactured in the following way...

Testing and evaluation

You will need to find out how successful your final design solution is by testing it with a group of potential users. How well does it match the design specification? How well have you worked? What would you do differently if you had another chance?

Testing and evaluation

As you work through your project, you will regularly carry out testing and evaluation. For example:

- Analysing and evaluating the research material you collected.
- Evaluating and testing existing products.
- Evaluating initial sketch ideas or samples and models in order to make the right decisions about which to develop further.
- Assessing the quality of your making as you go along.

Last of all, you must test and evaluate your final solution.

Providing the evidence

Make sure you provide evidence to show that you have practically **tested** out your final design in some way. Try to ensure that your findings relate directly to the statements in your original **design specification**.

Testing the final solution

To find out how successful your design is you will need to test it out. Some of the ways in which you might do this are by:

- trying it out **yourself**
- asking **other people** to use it
- asking **experts** what they think about it.

As well as recording people's thoughts, observations and opinions, try to obtain some data: how many times it worked; over what periods of time; within what performance limits; etc.

To help you decide what to test, look back to the statements in your design specification. Focus on the most important ones. If, for example, the specification stated that a three-year-old child must be able to operate your product, try and find out if they can. If it must be a colour which would appeal to young children, devise a way of finding out what age ranges it does appeal to.

in my design folder

What do you think you have learned through doing the project?

✓ Comparison of my final product specification with my design specification showed that...

✓ The people I showed my ideas (drawings and final product) to said...

✓ I was able to try my design out by...

✓ I discovered that...

✓ I could improve it by...

❝ *don't forget*

Try to identify a mixture of good and bad points about your final proposal and method of working. You will gain credit for being able to demonstrate that you are aware of weaknesses in what you have designed and the way that you have designed it.

How could you test these prototypes?

Evaluating the final solution

How successful is your final design? Compare it to your design specification. Comment on things such as:

- how well it works
- how well it meets the needs of the target market, and the requirements of the client, manufacturer and the retailer
- the appropriateness of the materials and processes
- what it looks and feels like
- what potential users said when testing it
- what experts said when assessing it
- whether it could be manufactured cheaply enough in quantity to make a profit
- the effective use of ICT to assist reproduction or manufacture.

Justify your evaluation by including references to what happened when you tested it.

Fast forward

Consider what you would do to develop or improve your product if you had more time. With reference to your evaluation, what aspects of the design would you try to improve?

■ Evaluating the process

How well have you worked?

- What has been the best thing about your project?
- How creative and innovative were your ideas?
- How good was your making?
- How effective was your use of ICT?

Suppose you were starting again. How would you improve the way you had researched, developed, planned and evaluated your project?

What have you learned as a result of undertaking this project? Have you managed to build on your strengths and develop your weaknesses?

If people have been critical of aspects of your design, do you agree with them? Explain your response.

Remember that evaluation is on-going. It should also appear throughout your project whenever decisions are made. Explain the reasons behind your actions.

Project presentation

The way you present your project work is extremely important. Remember you won't be there to explain it all when it's being assessed! You need to make it as easy as possible for an examiner and moderator to see and understand what you have done.

Tell me a story

All your investigation and development work will need to be handed in at the end, as well as what you have made. Your design folder needs to tell the story of your project. Each section should lead on from the previous one, show clearly what happened next, and explain why. Section titles and individual page titles can help considerably.

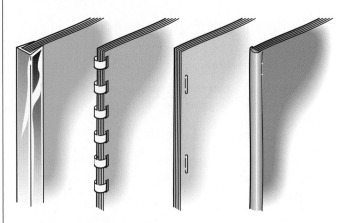

Binding methods.

" *don't forget*

Presentation is something you need to be thinking about throughout your project work.

Remember to include evidence of **ICT** work and other **key skills**. Carefully check through your folder and correct any spelling and punctuation mistakes.

■ Presentation checklist

There is no single way in which you must present your work, but the following suggestions are all highly recommended:

- Securely bind all the pages together in some way. Alternatively, use an inexpensive plastic folder.

- Add a cover with a title and an appropriate illustration.

- Make it clear which are the main sections.

- Add titles or running headings to each sheet to indicate what aspect of the design you were considering at that particular point in the project.

- Always work on standard-sized paper, either A3 or A4. Aim for about 20 sheets, packed with ideas and information.

- Aim to have a mixture of written text and visual illustrations on each sheet. You can use word processors and DTP to help create your sheets, but you will need to include hand-drawn sketches and notes as well.

- Try to get out of the habit of having to go back and re-do your project sheets 'neatly'.

- Include as many different types of graphic communication as possible.

- When using photographs, use a small amount of adhesive applied evenly all the way around the edge to secure them to your folder sheet.

- Think carefully about the lettering for titles, and don't just put them anywhere and anyhow. Don't make the size of titles and text any larger than necessary.

e-portfolios

You might be considering submitting your coursework entirely electronically. The main way of doing this is to use a presentation package. This can be useful if you want to include animation or video work. Remember, though, that these packages are not really intended for presenting project work for assessment. Think carefully about whether:

• what you hand in will be more convincing than if you work on paper

• it will take more or less time to prepare.

Make sure you discuss these points with your teacher.

Using ICT

ICT (Information and Communication Technology) is widely used in the design and manufacture of products. You can considerably enhance your GCSE coursework with the effective use of ICT – not just CAD-CAM, but also databases and spreadsheets, planning and presentation tools and, of course, the internet.

Using ICT in your project

The important thing is to decide when it is best to use a computer to help with your work. Sometimes it is easier to use ICT for parts of your coursework than to do it another way. On other occasions, it can be far easier and quicker to draw some sketches and make some notes on a piece of paper than use a computer.

The following are some ideas showing you how using ICT could enhance your coursework. You do not have to use all them, but try and use some!

Identifying the problem

- The **internet** can be used to search manufacturers' and retailers' **websites** for new products, indicating new product trends. Consumer-based sites might identify users' problems with existing solutions.
- 'Mind mapping' programs can help you identify and analyse the scope of the problem you are solving.

Investigation

- A questionnaire can be produced using a **word processor** or **DTP**. Results from a survey can be presented using a **spreadsheet** as a variety of graphs and charts.
- Use a **digital camera** to record visits and existing products.
- Visual images of the product, diagrams and other illustrations produced in a **graphics package** can be added to the presentation of your research and conclusions.
- A design or product specification can be written using a **word processor**.

going
on-line

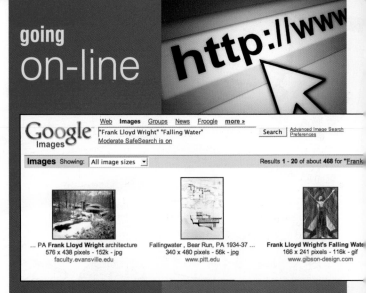

Finding what you want

To help you find the information you need on the internet you can use a search engine, such as Google.

Here are some useful tips.

- Phrase your search questions as answers. Try 'Frank Lloyd Wright was born in', instead of 'When was Frank Lloyd Wright born?'

- Put quotation marks around key phrases to reduce the number of results. For example, 'Frank Lloyd Wright architect' produces 28,000 instead of 5,390,000 references!

- To reduce the number further, use the 'Search within results' option. Entering 'Falling Water' (one of his most famous buildings), identifies just 281 sites.

- Remember to use the 'Image Search' – see how long it takes you to find a photograph of 'Falling Water'.

using
e-mail

Remember that text, photographs and computer files can be attached to e-mails and shared over the internet.

Some websites have e-mail addresses. You could try to contact experts to see if they could help with your coursework. It is important to be as specific as you possibly can, as these experts may be very busy people.

Become a blogger!

Post your thoughts, questions and design ideas on your own blog to get some responses from potential users and other designers. Specialist websites could be used to host questions for particular groups of people to respond to.

Developing ideas

- Ideas for your product could be produced using a graphics program, DTP or 3D design package. On screen they are easy and quick to change. Different shape, colour and surface pattern variations can be shown to the intended market for feedback. An increasing number of design packages are available on-line. These are often easier to use than professional packages.
- Likely costs of new products can be modelled using a spreadsheet. Different component costs can be varied quickly and easily, allowing you to see the consequences of changing your design ideas.

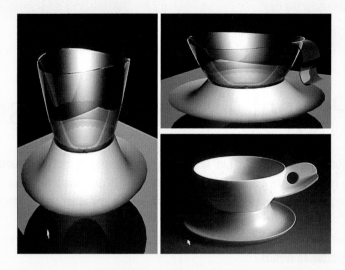

These cups were designed and rendered using a CAD program.

Project planning

A planning **time chart** program can be produced showing the duration of the project and what you hope to achieve at each stage.

Final ideas and manufacture

- Pre-programmed **CAM** equipment could be used to replicate manufacture (see page 10).
- Packaging nets can be produced using a **graphics package**. **Scanned** or **digital photographs** can then be placed within the nets to produce instant packaging designs.
- Parts lists for the costs of materials can be calculated and displayed using **spreadsheets**.
- A detailed flow production diagram could be produced using a **DTP** or **charting program**. Images could also be added to show important stages.
- **Digital images** can be used in the production plan as a guide to show how the product should be assembled, or to indicate its colour.

Product presentation

- Use **graphics packages** to prepare text and visual material for presentation panels. Charts showing numerical data can be quickly produced using a spreadsheet.
- Use a **presentation or multimedia package** to communicate the main features of your design, possibly incorporating animation, sound and video.

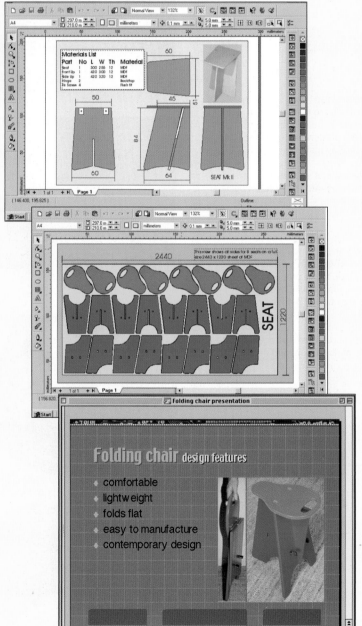

DESIGN & MAKE IT: PRODUCT DESIGN

| Remember the day? | p 72 | Where does it all come from? | p 74 | Where does it all come from? | p 76 |
| Surface pattern and texture | p 82 | Making a batch | p 84 | Getting the green light | p 85 |

Project 1: Sustainable souvenirs

'Sustainable Souvenirs Ltd' is setting up in business to make products from recycled or reused waste material. The company plans to produce a range of items that can be personalised for sale as souvenirs at museums and other tourist attractions.

Can you help the company develop some good design ideas for a range of suitable souvenirs?

About Us
The products that surround us have all been made using raw materials and energy that in many cases are finite — unless we are careful, one day we will simply run out of them. The manufacturing processes involved are often wasteful. When we have finished with a product, it is often just thrown away.

'Sustainable Souvenirs Ltd' employs experts who have innovative ways of recycling and reusing materials from products that have been discarded, such as crushed CDs, plastic water bottles and mobile phones.

Our Market
Tourism is a major industry. Wherever they go, tourists like to purchase souvenirs as a reminder of where they've been, or to give as gifts to friends or relatives. Often these products are standard items such as key-fobs, biscuits, pencil-cases, etc., with the name of the attraction and a picture of it added on.

Our Aim
We wish to become a major supplier of personalised souvenirs that are made from locally-available materials, reused and/or recycled products.

What you will need to do

- Decide what sorts of products make good souvenirs.
- Find out more about sustainable materials. What properties and characteristics do they have?
- Design a range of products that 'Sustainable Souvenirs Ltd' might be able to make.
- Finally, suggest how the products could be personalised for sale in batches to different tourist attractions.

Top and upper middle: **Plastic sheets made from recycled CDs and blue drinks containers.**

Lower middle: **Food produce is sustainable if it is grown or made locally.**

Right: **Many packaging materials are made from recycled paper and card.**

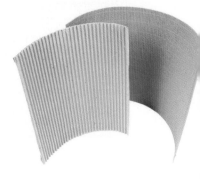

| Where does it all come from? | p 78 | Develop your design | p 80 |

What items are sold as souvenirs?

Souvenir badge from the 1951
Festival of Britain.

Souvenir keyring from
the Eiffel tower in Paris.

Souvenirs from various Disneyland theme parks.

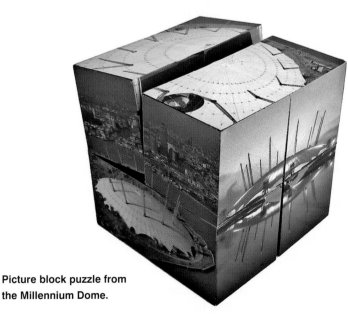

Picture block puzzle from
the Millennium Dome.

■ ACTIVITIES

1. Working in a group, brainstorm as many items as you can think of that could be sold as souvenirs. Remember different tourist attractions you have visited. Did you buy any souvenirs there? Make a note of all the different items as they occur to you.

2. Working on your own, organise the list of items into different groups or categories under the following headings according to what materials they are typically made from:

   • Wood, metal and/or plastic.
   • Paper and card.
   • Food.
   • Fibre.

3. Write down a list of tourist attractions in your area. Try to come up with as many as you can.

Remember the day?

How sustainable are existing souvenirs?
How well do they work as souvenirs?

Here's your chance to celebrate the good ones and complain about the bad ones!

 Before you complete this study, make sure you've studied 'Sustainable Product Design' on pages 12–15, and read 'Where does it all come from?' on pages 74–77.

Souvenir study

Study a number of souvenir items similar to those shown on page 71. In particular, consider their quality in terms of being an example of good sustainable design. You may have to adapt some of the questions to suit the particular product you are studying.

- Have any **reused** or **recycled** materials been used in its manufacture? If not, could they have been?
- Have the materials that have been used to make it come from a **local source**? If not, could they have been? Has the product had to be **transported** very far from where it was made?
- How **durable** is it (i.e. will it last a long time)? What part might fail first? If it were to break in some way, could it be **repaired**?
- Can it be taken apart easily or disposed of in some way to make it **reusable** or **recyclable**?
- Is the product **packaged**? If so, have sustainable materials been used? Does it require any batteries or any other type of power to make it work?

Next, consider other aspects of the **design** and **manufacture** of each souvenir.

- How well has it been '**personalised**' to the place where it is being sold?
- Who is the **target market** for the souvenir? In what ways might it particularly appeal to them?
- What do you like/dislike about the way it looks and feels?
- How well does it work as a reminder of the place it is intended to be a souvenir of?
- How well made is it?
- Would you buy it for yourself, a relative or friend? If so, why? If not, why not?

A brief history of souvenirs

For thousands of years, travellers have been returning with mementoes of the places they have visited. Sometimes these are natural objects that they have found or used tickets or programmes. Alternatively, they are purpose-made souvenirs.

The idea of the souvenir probably originates from Christian pilgrims around 2,000 years ago. They believed that physical contact with a holy object could transfer a spiritual blessing that could actually cure and protect them and their families. To avoid the relic being damaged, or even destroyed bit by bit, pilgrims began to be offered small, controlled amounts of oil, earth and water. These were supplied with special containers.

Another later solution was simple 'mass-produced' badges, cast in tin-lead, which represented the saint associated with the particular place of pilgrimage. During the fifteenth century, the market for expensive, high-quality souvenir products developed for wealthy merchants and travellers.

From the 1800s onwards, mementoes of visits to cities, buildings, figures and events started to become popular. These often took the form of illustrated papier mâché and wooden boxes, glass goblets and earthenware plates and mugs.

During the late nineteenth and early twentieth centuries, the working and middle classes had more money, and greater opportunities to travel. As a result, souvenirs became smaller and cheaper, and began to include seaside resorts and the First World War. This period also saw a large number of 'Great Exhibitions' of trade and industry across the world. These attracted millions of visitors and generated a demand for a large number of souvenirs.

Today, souvenirs are as popular as ever, and the available range of product types continues to expand. Examples of souvenirs from the past have become very collectable.

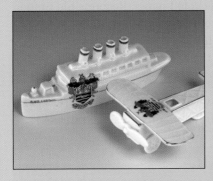

Porcelain models of the Lusitania, which was sunk in 1917, and a World War 1 warplane made in the early 1920's, both decorated with the arms and motto of Blackpool.

Souvenirs of the 1939-49 New York World's Fair.

An early religious souvenir.

■ ACTIVITIES

In groups, compare the souvenirs you have studied. Use the 'How does it score?' list on page 15.

a) State which souvenir you rate as the most sustainable and explain why.

b) Complain about the souvenir you think is the least sustainable and explain why. How might it be improved?

Where does it all come from?

Most of the raw materials we use come from different sources around the world. They usually need to be processed in some way before we can use them to make things. This is known as primary processing. Making the prepared materials into a recognisable product is called secondary processing.

Paper and board

Paper is made from a web of cellulose fibres, usually extracted from a blend of coniferous and deciduous trees. Water is added to tiny chips of wood to make wood pulp, which is poured over a fine mesh through which the water passes, leaving a flat layer of fibre. Waste paper, hemp, flax and cotton can also be used.

In industrial production, starch, clays and sizing agents are added to change the paper's surface properties and characteristics to produce a variety of textured cards and boards. Chemical dyes are also added to produce coloured surfaces.

Paper-making was invented in China and Korea over 2,000 years ago, but did not arrive in Europe until the end of the fifteenth century. The first paper to be milled in England was in Hertford in 1494.

Paper trail

Worldwide, there are approximately 10,000 paper and paperboard mills. About 300 million metric tons of paper and paperboard are produced each year. The United States produces nearly one-third of the world's total production. In Europe, most of our paper comes from Scandinavia.

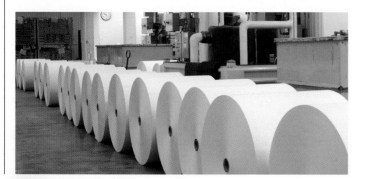

A sustainable industry

All this paper-making does not necessarily mean that we're running out of trees. Much of the raw material used to make paper comes from waste sawdust and wood chips, and from recycled paper. Responsible timber companies re-plant more trees than they cut down. However, large areas of forests are also destroyed each year causing damage to the ecosystem.

When timber companies do harvest trees for paper, they do not just cut them down and leave the land bare. That wouldn't make sense. Timber companies need trees to keep their companies in business.

■ **ACTIVITY**

Search the web for information about paper-making and the destruction of rainforests. Sites run by the paper-making industry will generally tell you that there isn't a sustainability problem, while environmental action groups tell a different story. Who do you think is correct?

Timber

Natural boards

Natural wood, or timber, is categorised as either softwood or hardwood.

- Softwood comes from trees that grow quite quickly: Piranha Pine is an example. These are usually conifers and are found in the colder areas of the northern hemisphere.
- Hardwood comes from deciduous trees that grow slowly, for example, Beech, Mahogany, etc. There are thousands of species of hardwoods across the world, including tropical regions.

After being cut down, the branches are removed and then the bark. The timber is then cut into planks or slabs. It then needs to be carefully dried or 'seasoned' to remove the excess water and moisture.

Manufactured boards

Manufactured boards are processed from particles, sheets or small pieces of wood. For example, to make medium density fibreboard (MDF) small particles of wood are mixed with glue, heated and compressed into flat sheets.

To make plywood, veneers or thin sheets of wood are glued together, heated and pressed, this is also known as lamination. Manufactured boards are very stable.

Sustainability matters

As with paper and card, it's important to ensure that timber is sourced from responsible suppliers. The Forest Stewardship Council monitors forest certification across the world and permits the use of its label where satisfactory standards are met.

Chemical treatments and finishes of timber are environmentally more serious than the use of the material itself. The use of preservatives and varnishes should be minimised if the product is for indoor use, or replaced with an organic coating. However, overall, timber is a better choice than plastic.

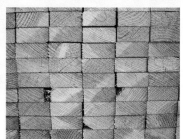

IN YOUR PROJECT

- Remember to specify that supplies of paper and card are only sourced from waste products and managed forests.
- Look out for opportunities to use 'defects' such as knots as design features.

Where does it all come from? [continued]

Metals

Metal ores are found all over the world, though some countries such as Australia, Brazil and India have more than others. The ores are mined or quarried and the unwanted earth, clay or rocks removed. They are then heated to a very high temperature to remove the oxygen and leave a pure metal that can be processed in a variety of ways.

All that glitters

Gold was the first metal to be discovered in 6000 BC, but at the end of the seventeenth century still only seven metals, including copper, silver, lead, tin and mercury, were known. Currently, there are 86 known metals.

Gold nuggets.

Sustainability matters

The mining of metals and their subsequent processing has a heavy impact on the environment and, therefore, their use should be reduced or avoided wherever possible. Aluminium and stainless steel are slightly more acceptable. However, metals can be effectively reused and recycled.

Plastic

Plastics exist in two forms – **natural** and **synthetic**.

- Natural sources include cellulose, latex and shellac, though these are rarely used in industry.
- The main synthetic source is **crude oil**. Crude oil is made naturally over millions of years from decaying sea plants and animals. The main suppliers of crude oil are in the Middle East. Crude oil is processed in a refinery where it is separated into various types, including petrol, diesel oil and naphtha, the latter being used for most plastics. There are a wide range of different types of plastics.

Fantastic plastic

The first plastic material was organic, and demonstrated in 1862. Celluloid, bakelite, rayon and cellophane were all developed around the turn of the twentieth century. PVC, Teflon and polyethene appeared during the 1930s. The widespread use of plastics in everyday products did not get underway until the 1950s however, but by the mid-1970s had become the most used material in the world.

Sustainability matters

Crude oil is a non-renewable resource – once it's gone, it's gone. Oil spills can cause substantial environmental damage, destroying wildlife, ecosystems and natural beauty. The production of plastic products produces greenhouse gases. Another problem with plastics is that when they are finished with, the majority do not degrade, i.e. break down, and are disposed of in landfill sites.

Manufacturers need to reduce or prevent the use of plastics and to reuse and recycle plastics as far as possible. To assist this, plastic products should be easy to take apart when thrown away. Another option is to use biodegradable and/or organic plastics, called Bioplastics. Polythene and PET are the most commonly recycled plastics.

Fibre

There are three types of textiles – **natural**, **regenerated** and **synthetic**.

- Natural fibres such as cotton, linen and silk are grown on plants, while wool comes from sheep. These fibres need to be cleaned and treated before they can be used in common products. Natural fibres are sourced from all over the world.
- Regenerated fibres, such as Viscose and Acetate are made from cellulose wood pulp, treated with chemicals.
- Synthetic fibres, such as Nylon, Polyester and Acrylic, are made entirely from petrochemicals, usually derived from oil or coal.

Fibres are spun into yarns which are then woven or knitted into fabrics. Dyes and other finishes are added to modify the surface and change its appearance.

Spinning a good yarn

The skill of weaving can be traced back 12,000 years to Neolithic times. Spinning spread from India and arrived in Europe in the late fourteenth century. The industry was mechanised during the nineteenth century during the English Industrial Revolution. Today, production is automated, and centred mainly in China.

Sustainability matters

The use of toxic chemicals during the treatment of natural fibres and the production of regenerated and synthetic fibres can cause extensive pollution and use large amounts of energy. Manufacturers need to start using organic or recycled cotton and wool, or specify reused fibres for other applications such as insulation. Water-based dyes are preferable to chemical ones, but vegetable dyes, or even un-dyed fabrics are a more sustainable solution.

Where does it all come from? [continued]

Food

Raw food materials come from **animals**, such as cows and sheep, and **plants**, such as wheat grain, sugar beet and fruit trees and bushes. These raw materials usually need to be processed into ingredients such as flour and refined sugar before they can be made into edible food products. This might involve milling, sorting, heat-treatment, refining and fermentation.

Raw food materials come from all over the world. Many are seasonal and need to be preserved in some way before or after processing, so that they can be made available all year round.

Sustainability matters

Until quite recently, the food we ate was produced locally. Today, the raw food materials are transported across the world for production somewhere in the UK. From there, they are transported again to the supermarket. It is the damage caused to the environment while being transported long distances that is the main concern in terms of future sustainability.

The phrase '**Food Miles**' refers to the distance in miles a food ingredient has had to travel from field to plate. This has been recognised as a good measure of sustainability.

The food production industry also requires large amounts of water and energy. Food packaging is also often very wasteful.

Manufacturers need to consider the development of food production at a more local level, and to focus on more environmentally-friendly forms of packaging.

FAIRTRADE

If a product displays the FAIRTRADE Mark, it means that farmers and workers benefit from high social and environmental standards, plus a premium that enables them to improve the lives of their families and communities.

Companies that buy direct from farmers at better prices, help to strengthen their organisations and offer consumers the opportunity to buy products that were bought on the basis of a fair trade.

Electronic components

Computer chips and transistors are made from silicon. Silicon is a commonly available non-metallic chemical compound that does not exist uncombined in nature, so it needs to be extracted from materials such as sand and other common rocks and minerals. A complex and sophisticated process of reduction and purification is needed to produce high-quality slices or 'wafers' that are free of defects.

Electronics is a relatively new industry. Transistors were first introduced in the late 1940s. **PCBs** (Printed Circuit Boards) were developed during the 1960s. Japan and other Pacific-rim countries lead the production of cheap and reliable electronic components.

Sustainability matters

Electronic products are typically complex and consist of a large number of components. Some contain quantities of precious metals, which, in environmental terms, are expensive to extract. When electronic products are disposed of, there can be problems with potentially hazardous substances (e.g. lead solder in older PCBs) that can enter the water or food system.

Radio frequency identification (RFID) chips being printed. They consist of a plastic circuit printed on to foil and antennae for receiving and transmitting radio waves. Plastic circuits are cheaper than silicon, although at the moment they are also slower and store less data.

IN YOUR PROJECT

- Try to develop products that can be more easily disassembled and recycled, and that contain fewer components.

- Avoid using manufacturing processes that create hazardous waste.

Ceramics, glass and concrete

Ceramics cover a range of products including chinaware, glass, concrete, bricks and tiles. Ceramic materials are made from three main materials: **clay**; **sand**; and a range of other **minerals**. Clay can be dug locally and worked by hand, but is generally developed on an industrial scale. Some ceramic products, such as reinforced concrete, have great strength, while others can be used at extremely high temperatures, e.g. heat-shield tiles on a spacecraft.

The first ceramic items are thought to have been made more than 24,000 years ago, but the range of products available today have only developed during the past 500 years and, in terms of the use of their electrical and thermal properties, during the last 50 years.

Sustainability matters

In comparison to many materials, ceramic products have a low environmental impact. Glass can be effectively recycled, depending on its type and what it has been used for. However, the production of chinaware tends to use water and other materials inefficiently and can generate a large amount of potentially hazardous and non-reusable waste.

Develop your design

Now that you know about the sorts of materials and processes that are sustainable,
you need to develop your designs for a range of products that 'Sustainable Souvenirs Ltd'
will be able to make.

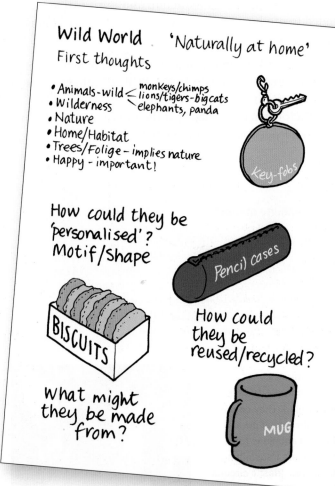

Design specification

Remember that your products:

- must be made from recycled materials, materials that can be reused and/or that are available locally
- must cause the minimum of damage to the environment during the manufacturing process
- need to be saleable as souvenirs
- need to be personalised in some way to each specific attraction.

The souvenirs could be something that can be used, worn or eaten. They need to be appropriate for historical visitor attractions such as museums, stately homes, castles and other 'heritage' sites.

Ideas for souvenirs: first thoughts

Refer back to the list of typical souvenirs you generated (see page 71). If you have not already done so, you need to finalise which items will be in your range of souvenirs. Maybe it could include two, three or more of the following:

- A key-fob containing a picture.
- A bookmark.
- Something in the shape of an animal.
- An edible product, suitably packaged.
- A picture frame.
- A badge or brooch.
- A set of drink mats (called 'coasters') or something else of your own choosing.

Eventually you will be making batches of your products, so try not to choose anything too ambitious that would use a lot of materials and involve more than a few construction processes.

Make sure the products you choose are suitable for development as sustainable products.

- What could they be made from? How could they be reused or recycled when they are finished with? How likely do you think it is that the process of their manufacture is environmentally friendly?
- How will each product be personalised to become a reminder of a visit to a specific attraction? Could it be printed on, or shaped in a particular way?

Try to think of ideas for completely different souvenirs – things you've never seen before but that might just catch on. Can you come up with something that is so novel that people would want to buy it just for that reason?

If I were you...

Discuss your ideas within your group. Listen carefully to the comments made by others. Have you got some good advice about someone else's ideas?

- How could their ideas be improved?
- Could the products be made in a more sustainable way?
- Are the products aimed at the right market?
- How good a reminder of a memorable day are the products?

Maybe some of the suggestions you made about someone else's design could be applied to yours?

Which are your best ideas and why? If you think you have some good ideas for a range of sustainable souvenirs, you can then go on to the next stage and start developing your design. If you're still not sure however, then go back to your list and consider some more possible ideas.

Second thoughts

When you have come up with an idea for your sustainable souvenir, you will need to consider what material to make it from, and how it will be made.

Materials and components

Depending on the materials and processes you are considering using, study:

- woods, metals and plastics on pages 24–27 and 75–77
- ceramics on page 79
- textiles on pages 28–29 and 77
- paper and board on pages 74 and 102–103
- electronics on pages 44–45 and 79
- food on page 78.

How to cut shapes? How to apply designs? Colouring? Safety? Durable? Sustainable?

more adaptable to any design

I will develop this idea, a key-fob offers good publicity and this design will be highly customisable

How to do whiskers?

Design or logo here

Wording here

LOGO HERE

Name of animal, place or organisation here. Could also be boys' or girls' names.

Think about how the working properties and performance characteristics of the materials might affect your design:

- How might you need to change it to make it stronger, lighter, more durable, etc?
- How might it need to change to make it easier to manufacture?
- What problems might there be?

Desired by design

How can you make the souvenirs as attractive as possible? You will need to explore their shape, form, colour, texture and pattern.

How will it be 'personalised' to a particular attraction? You will need to design a suitable motif – a simplified representation of some aspect of the attraction that can be added to all the products in the range.

Surface pattern and texture

As an alternative to a single motif applied to your products, you might want to consider adding pattern and texture to make the designs more interesting and desirable.

Pattern

Most pattern designs consist of one small design or motif that is repeated many times over the surface area of a material. The size of the individual design and the simplicity or complexity of the way in which it is repeated can make a great deal of difference to the way in which people respond to it. Over the centuries, many different ways have been developed to make simple and highly-complex decorative patterns.

The same motif can be repeated in different ways to create a different pattern. For example, straight, half-drop, mirror and joined repeats can be used. The same pattern can also be produced in different colourways so that several possibilities can be considered before a final choice is made, or the different alternatives produced.

Pattern-making techniques

Simple techniques for printing identical patterns over large surface areas include the following:

- **Block printing**: traditionally wooden blocks are used but blocks can be made from other materials. This is still a popular technique used in rural India today.
- **Stencilling**: shapes are cut from a piece of paper or vinyl, the stencil is placed on the fabric and paint is applied through the cut-outs. The Japanese invented this method and developed it into a fine art.
- **Screen printing**: this is developed from the stencil technique. The design is placed beneath the fine mesh of the screen, paint is applied on top of the screen and a squeegee is used to draw the ink across the screen.

CAD can be used to produce sophisticated patterns.

Industrial production of T-shirts using screen printing.

Texture: taking the rough with the smooth

All material surfaces have a texture. They range from very rough to very smooth. People react emotionally to texture in a similar way to colour. Textures can help add **contrast** or be **complementary**. Many different techniques have been developed over centuries to add a variety of textures to products.

Construction method

The way a material is made and finished often provides the basic texture. For example, weaving and knitting create fabrics with different textures. The combination of different weaves and yarns provides further variety.

Embellishment

Texture can be added to a surface by applying another material to it in some way. With fabrics, techniques used to do this include:

- **appliqué**: the application of one piece of fabric on to a background fabric
- **reverse appliqué**: the stitching of a motif or pattern on to layers of fabric. Part of the top layer is cut away to reveal the layer underneath.
- **quilting**: the placing of wadding between two layers of fabric and stitching a pattern on top
- **embroidery**: stitching coloured threads on to fabric to create a pattern.

Sometimes materials can be left with rough edges and surfaces to provide depth and be easier to hold and grip. At the other extreme, they can be made smooth and polished, and easier to clean.

The nature of the surface of a **mould** can transfer a rough or smooth texture to the final product.

Intricate embroidery designs traditionally created by hand can be easily produced using CAD-CAM.

Making a batch

'Sustainable Souvenirs Ltd' needs to know how to make your souvenir designs in large quantities.

All together now

When you've finalised your designs, you will need to make a final version of each one, adding a personalised motif and any pattern or texture to it.

After you've done this, you will need to think about how a small batch could be made. The number of products made will **depend** on exactly what the product is and the available time and materials you have.

You might be able to share the making with other members of your group – providing you help them make their batches of products in return!

For example, you might decide to make:

- 12 bookmarks made from laminated card
- 6 coasters made from plywood
- 10 shaped biscuits made from gingerbread
- 4 embroidered handkerchiefs.

This time, instead of making each item one by one, you need to work out ways to save some time by making them all together at the same time wherever possible.

Create custom artwork

transfer paper

PRINTER

Print onto transfer paper – large or small batches as required

Iron transfers onto base material

Transfers

4mm ply

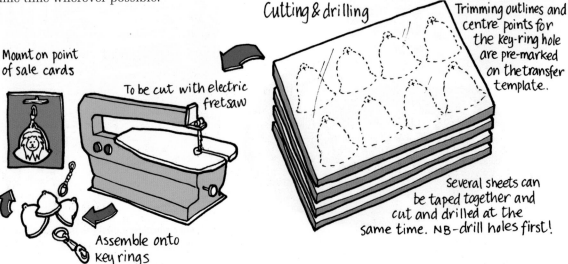

Cutting & drilling

Trimming outlines and centre points for the key-ring hole are pre-marked on the transfer template.

Several sheets can be taped together and cut and drilled at the same time. NB – drill holes first!

Mount on point of sale cards

To be cut with electric fretsaw

Assemble onto key rings

Getting the green light

You are about to send your ideas to 'Sustainable Souvenirs Ltd'. What would you tell them about your design? How could you convince them that it is a good idea worth investing in? Will they give it the green light to put into production?

Selling your design idea

Prepare an A4 or A3 sheet that includes a coloured drawing or photograph of your final design. What are the best features of your design? Highlight how it achieves each the following:

- How the product can be made from recycled materials and/or materials that can be reused.
- How the process of manufacture causes minimum damage to the environment.
- What makes it a good souvenir, aimed at a particular market.
- How it can be personalised in some way to each specific attraction.

Final evaluation

Ask the other people in your team to write a short evaluation of your final design and presentation for 'Sustainable Souvenirs Ltd'. You will need to write a short evaluation of theirs in return.

- Highlight the best aspects of their design.
- Comment on some of the things that could be improved.
- Suggest what they should do if they had more time.

Read the comments made by the other people in your group. Do you agree with them?

Finally, write your own evaluation, highlighting the best aspects of your work and the things that could be improved. Mention the points made by the rest of your team, and challenge their suggestions if you don't think they are correct.

| Product analysis and target markets | p 88 | Drawing and rendering systems | p 90 | Human factors: anthropometrics | p 92 |
| Any colour you like! | p 100 | Paper, packaging and print | p 102 | Packaging and point-of-sale displays | p 104 |

Project 2: Water works!

Can you design an innovative water container, together with its packaging and a point-of-sale display?

Water is good for you! If you are undertaking any form of physical exercise, it is especially important that you drink plenty of water.

DESIGNER DRINKS

Recently, there has been an increased demand for 'designer' sports drinks. These are bought by people who feel they need extra energy when exercising or participating in a local 'fun' event or competition. As a result, there is a wide range of brightly-coloured, vitamin-fortified 'designer water' available on the market. These attractive, and often expensive, bottled drinks claim to give the drinker vitality, focus and even stress relief.

WATER SPORTS

However, nutritionists advise that, although they might have some benefit for demand for professional athletes, sports drinks are of little value to the average user. Water, they say, is much better for you. Despite this, the 2012 London Olympic Games is likely to create an even bigger market for sports drinks in the future.

Article from a recent marketing magazine.

Ordinary water bottles do not make water look fun and exciting and energy-giving.

The task ahead

Rather than create a new sports drink gimmick to meet growing demand, a company called 'Water Works!' wants to develop a range of exciting, good-looking containers that people will simply fill with tap water or their own water-based fruit drinks.

Its shape, materials and colours will need to make drinking water an exciting and stimulating experience. It will also need to be easy to carry, hold, open and drink from in a variety of situations.

Can you come up with a range of designs for distinctive containers that 'Water Works!' can manufacture and market?

After a major sports event thousands of water bottles are just thrown away.

What you will need to do

- Study existing liquid containers to learn more about how they work; how they are made; and what the best solutions are.
- Develop ideas for unusual shapes and forms for water containers and carrying devices.
- Make a prototype of your product to show to 'Water Works!'

What you might also do

- Invent a name for the product and develop a memorable logo or symbol.
- Show how the name and visual identity could be displayed on the product and/or tag.
- Suggest a suitable way of packaging the product.
- Design and make an eye-catching point-of-sale stand to promote the product in sports shops.

Your teacher will tell you which of the above tasks you need to undertake to complete this project.

How many different situations can you think of in which people taking exercise need to take a drink with them?

■ ACTIVITIES

1. Working as a group, make a collection of containers and photographs of containers that hold liquid – not just for water or for sporting occasions, but for anything from teacups to perfume bottles and Tetra Paks ™ to paint pots.

2. Investigate the nutritional claims of 'designer' sports drinks. Conduct a search on the internet. Although nutritionists are not convinced, many others are. In your group, discuss whether or not these claims are justified, and whether or not users would be better off drinking water instead.

3. Sketch and make notes of your first ideas for a new and exciting water container. Share them with the rest of your group to get their response.

How well do each of these containers work as devices to:
- **carry and drink from**
- **open and refill?**

Product analysis and target markets

First, you need to study existing products that contain liquids. This will enable you to learn more about how they work, how they are made, and what the best solutions are. Think carefully about the sort of user they are intended for.

Which works best?

Your group will need to have at least four different types of empty liquid containers in front of you. Ensure that any potentially hazardous liquids have been thoroughly cleaned away first.

Make a quick coloured sketch drawing of each container. Leave some space around each drawing to add notes later.

Each of you takes turns to lead a discussion in your group of each container. Consider the following in particular:

- How do you know what sort of liquid it contains?
- What stops the liquid coming out when not required, or if it is knocked over?
- How easy is it to refill?
- How do you hold it to pour or drink from?
- Its shape, colours, graphics and/or decoration. Which words best describe how it looks and feels?

Cool?   Exciting?
Functional?
Useful?   Cheap?   Smart?

Also discuss:

- the characteristics of the liquids, e.g. hot/cold, thin/thick, edible/poisonous
- how much liquid it can contain
- what material(s) it is made from
- what colour(s) it is
- how heavy it is when empty.

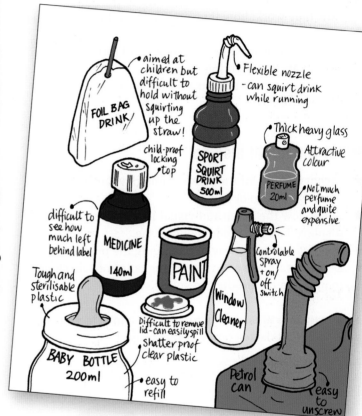

■ **ACTIVITIES**

1. After you have discussed all four containers, add some notes that record the most interesting features of each design. Don't just add descriptions (e.g. 'it is bright blue'), but say how well the feature contributes to the success or failure of the design (e.g. 'the bright blue colour successfully gives it a feeling of cold freshness'). Where you can, make comparisons between the containers.

2. Study the next page. Add notes to your drawings to explain the specific types of consumer each container is aimed at.

3. Conduct a survey of people who buy sports drinks. Try to discover why they buy them. Which are their favourite ones? Do they believe they really improve their performance? Write up your findings, including an illustrated graph or chart.

Target markets

Most products are aimed at specific types of people who are more likely than others to want to purchase one. It's important for you to decide who your **target market** is.

Everyone wants and likes different things. But certain groups of people tend to behave in similar ways. For example, many adults who drive particular types of car enjoy similar sorts of leisure activities. People who regularly eat convenience foods share a **lifestyle** that means they also buy similar magazines.

Consumer profiles

Market research organisations have come up with a range of ways of classifying lifestyle groups: they create **consumer profiles**. These provide details of different groups of people's lifestyles and buying habits.

Products and services are then aimed at particular lifestyle groups. Their design and advertising can be developed to make them appropriate to the environment and situations that those groups of people might most desire.

Finding the gap in the market

We buy things to satisfy our desires. We don't just buy a personal stereo, for example. We want one in a particular price range which will have a specific range of functions, and looks the way we want it to look to reflect our lifestyle.

Manufacturers produce a range of models to satisfy different markets. Companies are keen to spot a **gap in the market**, i.e. a product model or variation that is not well supplied by other manufacturers.

Consumer profiles

Wealthy achievers

Typically, these are males in their late twenties to early forties. Winning is important for them. Time is crucial to them as they are always busy. They travel first class and have expensive cars. At work they wear smart suits, and when exercising they can afford the latest in sportswear. They purchase hi-tech communication equipment and the latest gadgets.

Comfortably off

Many young, single people with good jobs have a high disposable income. They holiday regularly, usually abroad, and, although they are not afraid to party, they try to maintain a healthy lifestyle. They might well own a good-quality mountain bike. They regularly buy convenience foods from the more expensive supermarkets.

This lifestyle group wears 'designer' clothes and shops impulsively in the high street and using the internet.

Moderate means

Many traditional families with school-age children come into this category. They generally live in cheaper two or three-bedroom terraced houses. These people are careful with their money. They holiday in the UK or possibly Spain. They spend a lot on cable and satellite TV to view sport. They shop at the larger supermarkets where they buy basic food items and clothing. They enjoy undertaking sports and leisure activities together as a family.

Drawing and rendering systems

3D drawings range from tiny 'thumbnail sketches' on the back of an envelope to carefully-constructed images on a drawing board or computer screen.

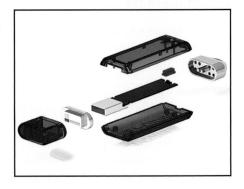

Exploded CAD drawing of a USB Datastick.

Isometric

Isometric drawings are used to show what an object looks like in 3D. They are quite quick to draw, especially if isometric grid paper is used as an underlay.

Planometric

A variation of isometric is called **planometric**. It is mainly used to help visualise room interiors or display stands. It differs in that the angles are drawn at 45° instead of 30°. This means that circular forms are easier to draw.

Perspective

Perspective sketches and drawings are similar to isometric drawings, except that the lines appear to meet at the horizon at 'vanishing points'. They are also more difficult to create.

There are two common types of perspective drawings: called 'one-point' and 'two-point'. The 'points' are the number of 'vanishing points' that the object recedes to.

- One-point perspectives are useful for views of simple objects or interiors.
- Two-point perspectives are useful for more complex products or room-interior drawings.

Cut-away drawings

When producing a **cut-away** drawing, the first decision is about the most suitable sections to leave out in order to show the detail required. Plans, elevations and isometric projections make a good starting point.

Exploded drawings

Exploded drawings are often used in instruction or repair manuals. They help to show clearly how a series of components fit and work together. An exploded drawing is usually shown in 3D. Isometric and planometric projections are often used for this purpose.

Working drawings

Working drawings are intended to enable someone other than the designer to make a product. For this reason, they have to be clear and accurate.

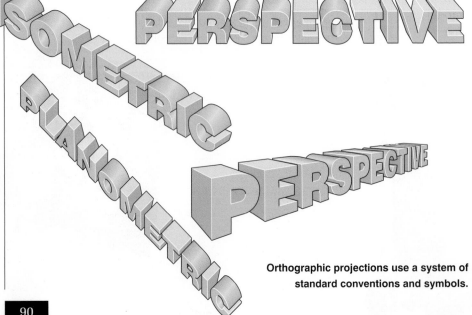

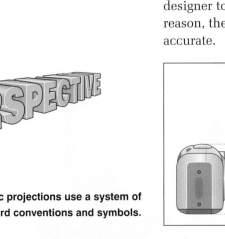

Orthographic projections use a system of standard conventions and symbols.

Making it look real

Designers use pencils, pens, chalks, paint and other tools to help make drawings of products look more realistic. Each media gives a different effect, and designers need to choose the one that communicates the shape, form and material being represented most effectively.

Making your mark

Different graphic marks can help show the surface texture of the material an object is made from, such as wood, metal or plastic. These can be shown to be smooth and shiny or perhaps hard and rough, or even reflective or transparent. Some drawings can be so realistic they look just like photographs.

This process of adding colour and texture to a drawing to make it look more like a material is called **rendering**.

Using lines

Using a different thickness of line can help make a form look more solid, or add emphasis to a particular component. Hatching can also be used to indicate different surfaces and materials.

Adding highlights and shading to indicate dark and light areas helps the representation of a 3D form.

On the computer

Drawings created using a 'Draw' or 'Paint' type program on a computer can be rendered in similar ways. Some programs have special filters and tools to create highly-sophisticated representations of form and texture. These are not easy to create, but can be quick to experiment with and adjust in order to achieve the effect required.

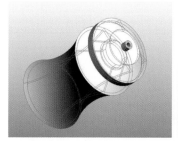
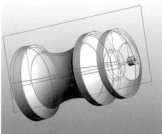
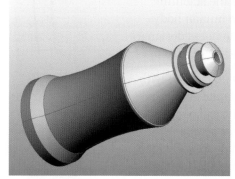

Pencils

Pencils can be graphite (black) or coloured. Hard pencils are used to create accurate plans and working drawings. Softer pencils are ideal for quick sketches. Variations of tone and texture can be easily created.

Felt markers

Felt markers are good for showing large flat areas of colour, particularly if the material being represented is smooth, such as plastic or metal. It is difficult to alter or remove a mark made with a felt marker.

Technical pens

Technical pens that produce fine ink lines are best for accurate plans and workshop drawings. They produce a much blacker line than a hard pencil, but are more difficult to correct.

Paints

Paints come in a wide variety of different types, from powder to water-colour, acrylic and oil-based. Water-based paints are less thick and can be more easily mixed and blended on a paper surface. Acrylic and oil-based paints are much brighter and suitable for applying to a wide range of hard surfaces.

Chalks and pastels

Chalks and pastels are quick and easy to use. They are particularly good at adding tone and shading as they can be quickly blended and graduated. They can also be used effectively on coloured paper or board backgrounds.

IN YOUR PROJECT

- Try to use simple cut-away and exploded drawings while sketching ideas.

- Also use them as part of final presentation work to help illustrate key aspects of your design proposals.

KEY POINTS

Rendering is used to:

- provide an impression of the 3D form of a product

- indicate materials and textures of surface finishes

- show colour.

Human factors: anthropometrics

**Your product needs to be just the right size for its target market –
not too big or too small to use comfortably and easily.**

What is anthropometrics?

Anthropometrics is the scientific measurement of the human body. Anthropometric data helps designers create products that are easy to use.

Anthropometrics is concerned with gathering statistical data about the dimensions and physical capabilities of different groups of people. Designers use the data for guidance, and it saves them having to collect large amounts of measurements for themselves.

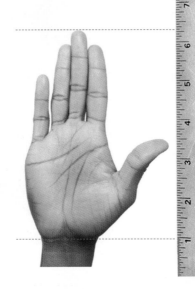

The data is not just about the overall height or width of people, but provides detailed analysis of all parts of the body, such as the size of hands and fingers. It also involves things like how far people can reach, the weight they can lift, the pressure they can exert, the extent to which the wrist can rotate, what angles of vision they have, etc.

Anthropometric data can be measured in two ways:

- **static**, i.e. when the body is still, e.g. sitting, standing or lying
- **dynamic**, i.e. when the body is moving, e.g. twisting, stretching, pressing.

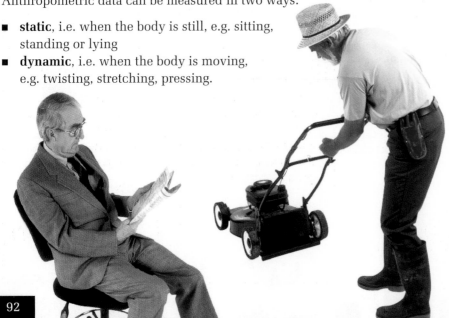

■ ACTIVITY

1. Study the photographs of people using various products on this page. For each one, discuss how anthropometric data might be used to help design the product. Suggest how the manufacturer has provided the possibility of adjustment to suit different users, and/or provides a range of models in differing sizes.

A calculated decision

In some situations a designer might commission a specialist company to collect accurate data for a specific situation not included in standard anthropometric tables.

For example, a design team needed to determine the best height for a push button on a calculator after it had been pressed. They wanted to reduce the amount of surface scratching caused by long finger nails. Measurements of the right forefinger nail lengths of a wide variety of the population were taken to the nearest 0.5mm.

The results revealed that the push button would need to be 4mm above the surface to avoid all scratching. However, only 5% of people had a nail length of more that 3mm, so 3mm was used as an acceptable height.

Different, but similar

Anthropometric data is subdivided into many different categories. For example, British males aged 20–29, British females aged 20–29, Japanese males aged 50–59, and so on. Sizes can vary widely according to age, gender and racial origin. Other factors such as diet, injury or disability can also have an effect.

Designing for Mr Average?

Using anthropometric data is more than just using averages of the population. The number of people who are exactly average is very small. The data is mainly used to identify the extremes of size. Generally, designers aim to accommodate 90% of likely users. These means that a product can be used by all but the smallest 5% and the largest 5% of the target market.

Anthropometric data is usually published in table form, showing the measurements for different sexes, ages and races at the 50th percentile (i.e. the 'average'), and the 5th and 95th percentiles.

Gripping stuff

Designers make detailed studies of how people hold and operate different products.

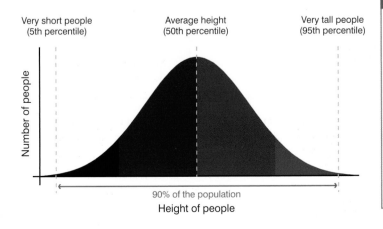

Very short people (5th percentile) — Average height (50th percentile) — Very tall people (95th percentile)

Number of people

90% of the population
Height of people

IN YOUR PROJECT

As you develop your bottle design idea, you will need to consider the overall shape, the hand grip and the method of opening and closing the container.

KEY POINTS

- Anthropometric data concerns the physical dimensions and capabilities of human beings.

- Ergonomics is the study of the way people interact with products in their living and working environment.

Going mouldy

The way in which your new product is going to be manufactured will affect the size, shape and form it can be.

Many familiar everyday products are made in moulds. Using a mould enables complex curved shapes to be reproduced quickly, accurately and in large quantities.

Making a mould

Moulds are widely used in the manufacturing industry. They enable identically-shaped products to be reproduced accurately and quickly.

The first step is usually to make the shape required in solid form in wood or some other resistant material. This is called a **former**. The sides of the former are then used to make the mould.

A gelatin mould for producing astronaut-shaped chocolates.

Blow-moulding

Most plastic bottles are manufactured using a process called **blow-moulding**. Blow-moulding is a common type of plastic moulding, often used to produce plastic vessels and containers.

Other types of moulding processes

Other methods used in industry to form plastics include:

- **vacuum forming** – a hot thermoplastic sheet is placed over a wooden former and the air sucked out to make the cooling plastic take on the shape of the mould
- **injection moulding** – plastic granules are melted and pushed into a metal mould
- **extrusion** – plastic granules are melted and pushed through a die to make tubes, rods, pipes, etc.

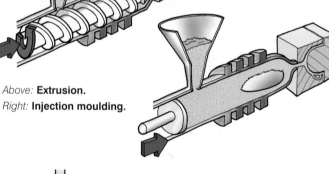

Above: **Extrusion.**
Right: **Injection moulding.**

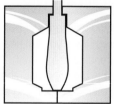

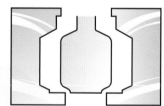

Plastic granules are fed in through a hopper. Inside, a motor turns a thread that pushes the granules though a heated section to turn them into a liquid state.

The plastic pours into the mould in the form of a tube, called a parison. Air is forced in making the sides of the tube cling to the sides of the mould cavity, forming a hollow object the same size and shape as the inside of the mould.

The plastic hardens on contact with the cold mould. The mould is then opened and the bottle released.

Case Study ■ SIGG

Not all liquid containers are moulded of plastic. Some are extruded in aluminium.

History of SIGG

In 1908, Küng, Sigg & Cie started the manufacture of leisure goods, kitchenware and electrical appliances. Since 1998, SIGG has concentrated on drinking bottles and is now a famous global brand leader for high-quality drinking systems for leisure and sports use.

Production

SIGG bottles are extrusion-pressed from a one-piece aluminium blank, which results in uniform, seamless walls.

The starting point is a chunk of aluminium the size of a hockey puck. This is extruded into a cylinder and pressed into any number of novel shapes. It takes about 26 manufacturing steps to form the neck.

The threaded ring is inserted and secured and the SIGG bottle is thoroughly cleaned inside and out, using ecologically-friendly cleaning agents. The bottle interior is sprayed with a taste-inert, food compatible stove enamel which is baked on.

Next, comes the outside coating. A solvent-free powder paint is applied and heat-bonded with the underlying metal.

The logo is now printed, along with the graphic design, then the leak-proof bottle top. Any SIGG bottle top fits any SIGG bottle.

There is a special range of SIGG bottles aimed at children, and a variety of carrying devices.

First thoughts

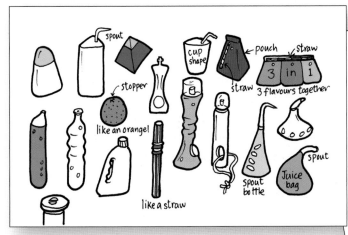

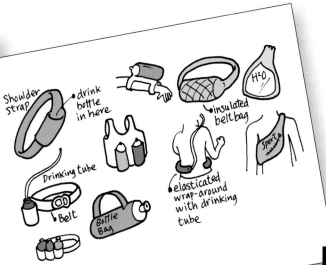

Water works!

What's in a name?

Brand names identify the company who makes the product. Product names identify a specific model or type. Brand and product names need to be chosen carefully as they say a lot about the nature and quality of the manufacturer and the product. It's also very important to choose the right style of lettering to match the sounds of the words and reflect the visual identity of the organisation.

Did you know that the word Oxfam is a shortened form of 'The OXford Committee for FAMine relief', the original name of the organisation founded in 1942?

Naming names

Brand and **product names** are important. They provide instant clues to the particular qualities of the product. Sometimes people are influenced into buying something simply because they like the sound of its name. Product names can reinforce the idea that a product is perhaps new, scientific, environmentally friendly, traditional, masculine, feminine, luxurious, efficient, good value, and so on.

Brand loyalty is important to companies as people sometimes buy a product because they are confident that it will be similar to the quality of previous products they have purchased.

Short and sweet

Brand and product names need to be easy to read, say and remember. Short words are popular – OXO, Next and Kodak, for example. Names that can also be easily understood and spoken in other languages can help global sales.

Words that sound like the product can be effective. Repeated or rhyming words help people remember a name. Fun names are often popular, especially when they include a clever play on words. Something called 'Fiesta' sounds bright and frivolous, while a 'Carlton' sounds managerial, classic and superior.

Some brands become so successful that their name is used to describe all similar makes. For example, 'Hoover' is used by most people when they refer to a vacuum cleaner.

So what are you going to call the product you are designing?

Making a Name for Yourself

The easy part of brand naming is finding something short, easy to pronounce and memorable. The hard and expensive part is avoiding unfortunate different meanings of the name in different languages, and making sure it can be registered as a trademark.

Getting the right associations is the next step. Sainsbury's soft drink 'Gio' has a get up and go, holiday, beachy feel to it. It called its cola 'Classic' because it wanted to send the message that it is as good as anything else on the market.

But Persil and Ariel tell us nothing about soaps. We accept without question that Typhoo is a tea, Anchor a butter, Apple a computer, Orange a mobile phone. If someone offered you a food product called Frog's Nose, you would probably be disgusted – but you don't worry about eating Bird's Eye foods!

Alan Mitchell

■ **ACTIVITIES**

1. Make a list of well-known product names. Identify how they work. What do they communicate about the nature and quality of the product?

2. Make a list of all the words you can think of that have something to do with the design of your product.

   - What does it look like?
   - What does it feel like?
   - What does it taste like?
   - What does it smell like?
   - What sounds does it make?

   Then, combine some of your words together to see if you can transform them into something interesting that will be appropriate to your target market.

Take a letter

How words are printed makes a great deal of difference to how we respond to them. **Typography** is not just about neat lettering. The appearance of words can be used to attract attention, look decorative and interesting, and suggest a particular time or place. Designers need to select typefaces very carefully. They need to specify a particular font, its size and weight, its colour and the amount of space around it.

Talking type

Each style of lettering has its own very distinctive tone of voice. Different typefaces can express a wide variety of feelings. It is possible to use a typestyle to help shout or whisper a message, or to attract attention to it. Alternatively, you might want your words to look and sound:

- traditional and reassuring
- romantic and relaxed
- stimulating and exciting
- creative and unusual.

Some typeface styles make us think of particular situations or periods of history.

Choosing the weight of a typeface can make a great deal of difference to its appearance.

Courier can make it `Official` Times Roman speaks with Authority Folio Bold Condensed expresses **Doubt?** Mistral shouts *Surprise!* Kabel Medium smiles with JOY Futura Extra Bold Condensed screams **Anger!**	Extra Light Light Medium *Light italic* Light condensed *Medium italic* Medium condensed Medium extended Medium Outlined **Bold** ***Bold Italic*** **Bold Condensed**

Legibility

It is essential that typefaces are easy to read. Some typestyles are more legible than others, but their size and the space between the words and letters makes a great deal of difference.

 Serif – easy to read, looks traditional.

 San serif – strong, bold and clear. Modern looking. Often used for titles and headings.

 Script – looks more personal and, depending on the style used, historical. Can be difficult to read.

Decorative – attracts attention and gives text a particular feel or association. Can be difficult to read. Best used for main titles.

■ ACTIVITIES

3. For each of the following words, choose an appropriate style of lettering. Write each word out in the style you have chosen or use a word-processing, DTP (desktop publishing) or graphics program.

   Goal!   Cool   Energy

   a) Choose a colour for the lettering and a different colour for the background.
   b) Experiment with different shapes for the background.
   c) Try adding texture and further colours to achieve different effects.
   d) Choose other words of your own to design. Discuss your work with others in your group. Make notes to evaluate which you think are the most successful and explain why.

4. Finalise a design for the name and style of lettering for your product.

Logos and symbols

Most companies and products use a visual symbol or distinctive logotype to enable the public to rapidly identify its name. The best symbols are simple and distinctive, and also manage to say something about the nature and quality of the products or services on offer.

Identify yourself

Most graphic identities are based on one or a mixture of the following:

- A design which uses the initial letters of the name of the organisation (known as a **logogram**).
- A design in which the name of the company is written out in full in a specially-designed and highly-distinctive typeface (known as a **logotype**).
- A design which is a simplified illustration of the product or service the company offers (known as a **symbol**).
- A design based on a decorative shape, pattern or form (also known as a symbol).

Colour also plays an important part in the design of visual symbols (see pages 100–101).

■ ACTIVITIES

1. Look carefully at the company and product logos and symbols on this page. What approaches to design is each based on?

2. Make your own collection of symbols and logos. State how effective you think each is.

   - Is it easy to identify?
   - What does it tell you about the company?
   - Why have its colours been chosen?
   - How would you describe any typefaces used?

Thirsty work!

'Water Works!' require a distinctive logo or symbol for the product name of the new water container you have designed.

- Use the product name you created on page 96.
- Experiment with different typefaces, sizes and graphic devices. Ideally, use a computer graphics program.
- Think about the shapes and colours that will appeal to your target lifestyle market.
- How does your design fit with the shape of your bottle? Where might it be placed on the bottle?

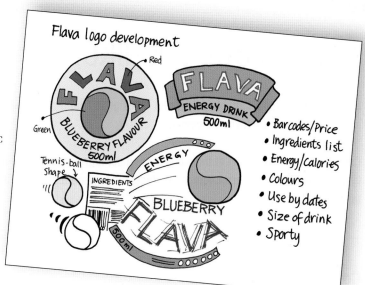

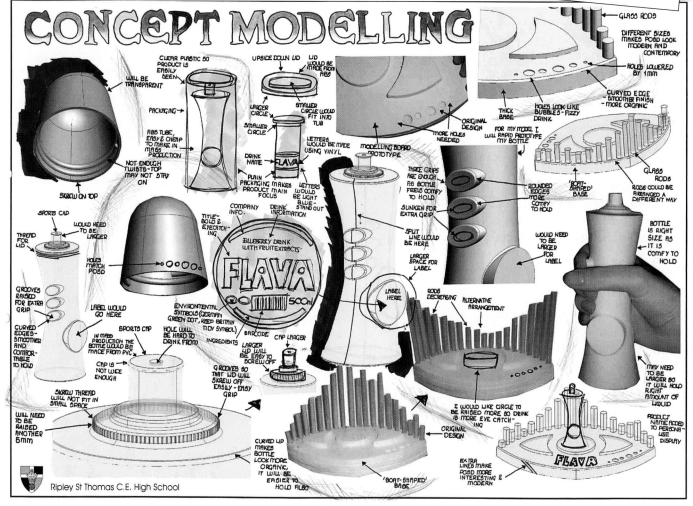

Finalise your design

Now is the time to start to bring all your ideas together – the shape of the bottle, the cap, the logo, packaging and a point-of-sale display. Work through pages 100–107 first.

On your development sheets. aim to use sketches, CAD drawings, photographs of soft models and 3D prototypes to explore and explain your design. Always keep in mind that you need to meet the needs of your target market.

Any colour you like!

Deciding on the most suitable colours for a design is a lot more complex than just picking your own favourite colour.

Colour is an essential part of the world around us. It provides us with important information about our surroundings. It affects the way we feel about, and react to, everyday things.

Colour and shape

Combining certain shapes and colours can produce very powerful visual images. A bright yellow bolt of lightning clearly serves as a warning. The soft shape of the cloud, combined with its warm, grey colour is much less threatening. Look for examples of how road signs and company symbols combine shape with colour.

Red is the colour of fire and suggests things such as strength, love, danger and disaster and evil. In the dark, red light produces the most easily visible contrast. Red is a warm and inviting dominant colour.

Colour associations

From very early times, colours have been connected in the mind with feelings, such as those of danger or happiness. Different societies and religions across the world use different colours to represent similar ideas.

Designing with colour

To create a colour scheme for a product, space or printed surface, experiment by choosing a range of harmonious hues and tones and add a small amount of one complementary colour. Also keep the following in mind:

- Warm colours seem to be closer to a viewer than cold colours, which seem further away. This also has the effect of making a warm-coloured object look larger than a cold-coloured object, even though they may both be the same size.
- Colours and tones can also be used to make an object look heavier or lighter.

Corporate colours

Many companies use a particular colour to help establish their brand identity. Different colours are more suited to different types of products and services.

- **Red** is often used to whet the appetite and to sell fast food. **Pink** is associated with femininity.
- **Orange** suggests power and is used for strong cleaners, health foods and drinks.
- **Yellow** represents long life (i.e. durability), sunlight, and sunsets.
- **Green** is the colour of nature and safety and is used to indicate 'environmentally-friendly' products. It is also the colour associated with money.
- **Blue** stands for reliability. For this reason it is popular with banks, travel and insurance companies, and international corporations.
- **Purple** expresses smoothness and excitement and is often used for luxury, and exotic products.

Primary colours can be mixed together to form Secondary colours. A Primary colour mixed with a Secondary colour will produce a Tertiary colour.

■ ACTIVITIES

1. Working as a group, make a collection of packages, logos, advertisements, etc. that predominantly use one of the colours mentioned. Create your own rainbow display.

2. You have been asked to suggest a colour scheme for the packaging of a new flavoured milkshake drink. For each of the following flavours, suggest *two* harmonious colours and *one* contrasting colour to be used on the outside of the container.

   a) Chocolate.
   b) Strawberry.
   c) Banana.

IN YOUR PROJECT

- Experiment with different colour schemes.

- Think carefully about the things people might associate with the colours you choose.

Orange is commonly associated with warmth and cheerfulness, sunsets and the autumn.

It is a very intense colour however, and is best used sparingly and to indicate hazards.

Yellow reminds us of the sun and in some religions symbolises life and truth. Yellow is also the colour of cowardice however. Yellow and black provide the greatest contrast and are, therefore, often used to indicate hazards. Yellow is a highly stimulating colour, so needs to be used sparingly.

Green is symbolic of the natural world. When used with blue, green suggests ice and coldness. With reds and oranges, it takes on the sense of autumnal, earthy scenes. Green is used to indicate first aid and safety. It often represents growth and hope. Generally, green is a calm colour.

Violet and purple are very rich colours, suggesting wealth and extravagance. They often denote royalty and can also indicate knowledge, nostalgia and old age. Purple and violet in large amounts are unsuitable for environmental settings.

Blue is the colour of the sky and the sea, suggesting far horizons and vast depths. It often implies things like truth, peace, loyalty and wisdom. Blue is used to identify electrical equipment. Blues are easy to live with, pacifying and calming colours.

Black is traditionally the colour of death and mourning in western civilisations. Witches are black. Black and grey clothes suggest uniformity.

White is the colour of snow and the moon. It represents purity and innocence and is used for christenings and weddings. White suggests cleanliness, which is why many washing machines and detergents are white.

Paper, packaging and print

Paper and card, or board, are used extensively in the packaging industry. There are many types to choose from. All of them come in ranges of size, weight, thickness, colour and finish.

Packaging and other paper and card products are usually printed using one of a number of methods.

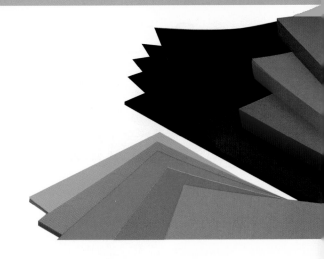

Size

There are many sizes of paper, card and board, but the most widely used is the 'A' series. This ranges from the largest – A0, to the smallest – A6. Each size is half the size of the previous one.

Weight

There are also many weights, or thicknesses, of paper. The unit of measurement used is grams per square metre (known as gsm). A typical sheet of exercise paper is about 70gsm, while a good quality magazine cover may be 150gsm (i.e. just over twice as thick).

Thicker card and boards are sometimes sold in units called microns. A micron is one millionth of a metre. A standard card used for packaging might be around 350 microns.

Treated papers

During manufacture, most paper, card and board is treated in some way to make it more suitable for a specific purpose:

- **Sizing** agents improve water resistance.
- More **bleach** can be added for extra whiteness, or the paper can be passed through highly-polished rollers for a glossy appearance.
- **Dyes** provide colour and different mixtures of pulp create textured effects. Some papers and boards are made with higher proportions of recycled paper.
- Other papers have special finishes to make them more suitable for different printing applications, such as colour photos.

Paper can also be classified in terms of its:

- durability – how long it lasts
- brightness – how well it reflects light
- texture – how rough or smooth it is
- opacity – how transparent it is.

Material	Applications	Characteristics
CARDBOARD (300 microns)	• High-volume colour printed cartons	• Low strength/weight ratio • Excellent printing surface • Recyclable
CORRUGATED CARD	• Protective packaging for fragile goods • Low-cost protection	• Recyclable

Material	Applications	Advantages
DUPLEX BOARDS: (pure wood pulp) unbleached body plus a bleached liner (350–640 microns)	• Tobacco, food and pharmaceutical packages • Provides textured printing surface	• Good for printing high-speed automatically-packed cartons
SOLID WHITE BOARDS: pure bleached wood pulp (350–640 microns)	• Book covers, cosmetics cartons	• Very strong • Excellent printing surface
MEDIUM DENSITY FIBREBOARD: high content of recycled paper and board (2–50mm)		• Lower cost • Higher strength/weight ratio • Can be lined • Good for long-run printed graphics
CAST-COATED BOARDS: a heavier and smoother coating applied to duplex and solid white board	• Luxury products with expensive-looking decorative effects	• Gives higher gloss after varnishing
FOIL LINED BOARDS: foil can be laminated to all the above board types (350 microns)	• Cosmetics cartons, pre-packed food packages	• Can be matt or gloss, gold or silver • Strong visual impact • Provides a barrier

Other packaging materials

Glass

Glass packaging is used mainly for liquids such as drinks and perfume, and for jams and other food products with a liquid content.

Metals

Aluminium and tin plate are used to make cans. Graphics are usually printed on to a label that is wrapped around and stuck on the can. For special effects, it is also possible to print directly on to the tin.

Plastics

Plastics can be made clear or opaque in a variety of colours. It is cheaper and lighter than glass, and easier to form into complex and unusual shapes and textured surfaces. Heat and pressure are applied to the plastic to produce the shapes required. There are many different types of plastic. They are usually called by their initials:

- LDPE: low-density polyethylene.
- HDPE: high-density polyethylene.
- PP: polypropylene.
- PS: polystyrene.
- PET: polyethylene terephthalate (polyester).
- PVC: polyvinyl chloride.

Commercial printing methods

There are four main methods of printing:

- Lithography.
- Gravure.
- Screen printing.
- Flexography.

The method chosen depends on the length of the print run, the quality required and the cost.

■ **ACTIVITY**

Match the following graphic products with the most suitable industrial printing method. Explain why your chosen printing method is the most suitable.

a) A school textbook.
b) The graphics on a plastic bottle.
c) A reproduction of a work of art as a poster.

Lithography

In lithography a plate is made that has areas that attract ink and other areas that repel ink. This is a popular method of printing that is economical for medium and long print runs of magazines, posters and packaging.

Screen printing

Screen printing uses a stencil through which ink is forced. This method is low-cost and only suitable for short print runs.

Fine detail is not possible, but it is effective at producing bright colours and bold shapes. It can be used for posters, T-shirts and large shop display boards.

Gravure

Gravure involves the production of an etched plate on to which ink is poured. It is used for very high-quality long-run print work, such as stamps or expensive magazines and illustrated books.

Flexography

Flexography uses a plastic or rubber relief plate. This method is used for printing on to unusual surfaces such as plastic bags, corrugated card and wallpaper.

IN YOUR PROJECT

Choose which type of paper or board and method of printing is the most suitable for the job it has to do.

CR 439 0-4-4T. Bo'ness and Kinneil Railway, West Lothian

GCR 8K 2-8-0. Great Central Railway, Leicestershire

Packaging and point-of-sale displays

Packaging design is an essential part of the manufacturing and distribution process. Meanwhile, point-of-sale displays help catch the attention of the target market.

Packaging matters

Packaging helps to ensure that the products being made end up arriving in the retailer's shop or warehouse and the consumer's home or workplace in the same condition in which they left the factory.

Products need to be contained so that they:

- are **protected** from damage, theft or contamination
- can be quickly **identified**
- are easy to **carry** and **transport**
- can be stored **safely** until needed.

Packaging serves to contain things that are:

- **perishable**, such as food and drink
- **dangerous**, such as chemicals
- **precious**, such as jewellery
- **numerous**, such as the parts of flat-pack, self-assembly furniture
- **liquid**, such as shampoo, fragrances, motor oils.

Packaging design

Packaging is often used to help sell the product. Eye-catching graphics need to draw people's attention to the product on the shelves and make it look desirable. At the same time, it must not give a misleading impression of the contents of the package.

Levels of packaging

There are three main levels of packaging: primary; secondary; and tertiary.

Primary packaging

This type of packaging is used to protect the product from damage and pilfering, and to provide information about handling, storage and the enclosed contents.

For example, a brown corrugated cardboard box containing one large item such as a TV, stereo system or refrigerator, or several smaller boxed items such as tinned goods, cornflakes, bottles, etc.

Secondary packaging

This describes the immediate package containing the product. This is usually made of thinner card of a better quality. It is frequently highly decorated, showing the item and giving detailed information to the purchaser, e.g. legal requirements, assembly instructions, safety warnings.

Tertiary packaging

This describes any final protective covering, such as sweet wrappings, plastic bags holding small components.

Getting to the point

New products are often promoted using a **point-of-sale** display structure. Such displays need to be simple and cheap to make, as they will not be used for long. Bold, bright shapes and colours are needed to make them stand out and catch the shopper's eye from a distance.

The term 'point-of-sale' refers to the place in which a customer will decide to purchase the product – not just the shop, but the specific areas in which products are displayed and where money and goods are exchanged – often a counter, or sales desk.

Newly-launched products, or those with a special promotional or sale-price offer, are promoted by means of a special display. The displays are only likely to be needed for a limited period (e.g. three weeks) in all the retailer's stores, which might typically range from half-a-dozen shops to several hundred.

Production requirements

The **batch production** of point-of-sale displays is an essential consideration. The numbers required will be limited and, therefore, they will be relatively expensive to produce. Some may even be 'one-offs', perhaps made by hand. Some may need to be made from card, or use more expensive materials, such as acrylic sheet, wood or metal.

They may have to be designed and made to last a long time or be reused, or they may only be required for a day.

Suitable printing processes need to be identified. Screen printing is suitable for short production runs and for printing a range of different materials, and produces bright and bold shapes and colours very effectively.

Design guidelines

A point-of-sale display might simply be an arrangement of the products themselves, possibly in a specially-designed container or dispenser.

Often these displays are interesting cardboard shapes involving complex nets, folds and cut-outs. They might also involve posters, 3D items, stickers on the window and special signs.

Correct positioning in the store is essential to ensure that it is noticed and seen by potential customers. It could be at a suitable height at the entrance to a shop, or on the forecourt of a garage, or wherever people are most likely to pass or linger. To help attract attention it can be brightly lit, or even musical!

Point-of-sale displays have to survive in quite hostile situations; they have to resist the everyday rigours of life in a shop, bright sunlight in the window, dust, etc. They have to be easy to install, keep clean and restock as appropriate.

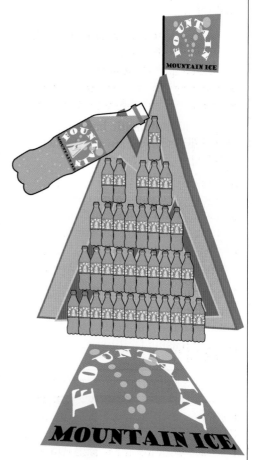

Surface developments / Final evaluation

Surface developments (nets)

Point-of-sale display stands are often made from stiff, flat card or board that is cut, folded and slotted together. The retailer often receives a point-of-sale display unfolded. This makes it easier and cheaper for the manufacturer to transport. The shape of the card before it is folded up is called a **surface development** (or **net**).

Boxes and bases

Tuck-in boxes are those that include flaps and slots. When industrially made, the corners of the tabs and flaps are radiused to smooth out the closing action.

Boxes with **automatic bases** slot together easily to speed up assembly. They are also known as 'crash-lock' bases.

Using CAD-CAM

Nets can be created very effectively using CAD-CAM. A package can be designed on screen in 2D. The computer can then simulate what the net will look like when folded up into 3D. The packaging graphics can also be applied to the 3D model.

When the net has been finalised, its dimensions can be sent directly to a machine at the manufacturer's that will cut the shape out, ready for folding. Some CAM machines have a creasing facility as well as a cutting tool.

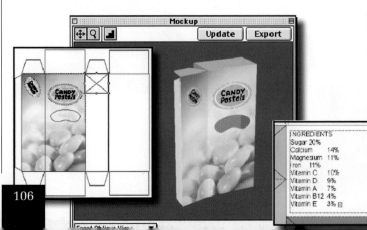

Cutting nets

In industry, nets are cut out using a die-cutter. This is a machine that stamps out the nets in a similar way to pastry cutting.

A special blade, known as a cutting rule, is cut, bent and shaped to fit into a laser-cut plywood frame. There are also creasing rules (which have a rounded edge and simply press into the card to assist with folding), perforated rules (for tear-off portions such as price tags) and a whole range of specialist blades (such as those used for jigsaw puzzles).

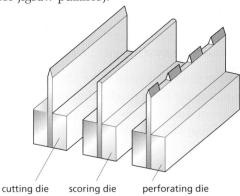

cutting die   scoring die   perforating die

The blade is surrounded with a stiff foam material to help to push the card off the blade once it has cut through the card. This die-cutting tool is sometimes called a press-knife but in the packaging industry is generally known as a **cutting forme**.

Formes are designed so that they can cut the maximum number of nets from a single sheet of card. This process is called nesting and a similar process is used in many different industries to maximise the sheet material and to reduce waste.

FINAL DESIGN

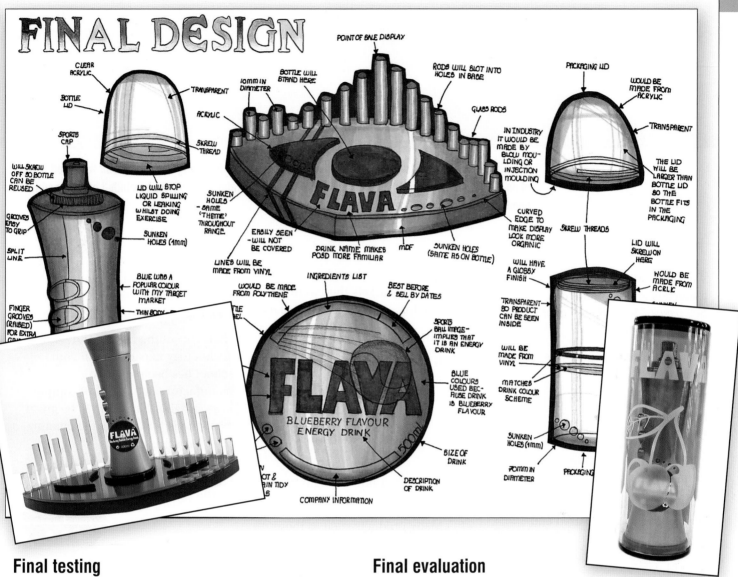

Final testing

Test and evaluate your point-of-sale unit.

- How long does it take to assemble the unit?

Place the product, or products, in the display unit.

- Does it hold them securely?

Take the unit along to an appropriate location and ask if you can place it on the counter, or at some other suitable place for a short period of time.

- How well can it be seen?
- Does it attract the attention of passers-by?
- Does it appear to be strong and durable enough?

Take photographs of the unit in use. If possible, obtain and record the opinions of the shop manager and some customers from the target market.

Final evaluation

Write up an evaluation of your project as a whole.

- Did you complete all of the things you were asked to? If not, why not?
- What were the best things about your work? What could have been improved?
- What have you learned while doing this project?
- Study some existing water bottles that have unusual shapes. How do they compare and contrast with your proposed design?

Q1 Examination questions

You should spend about one and a half hours answering the following questions. You will need some plain A4 paper, basic drawing equipment and colouring materials. You are reminded of the need for good English and clear presentation in your answers.

1. This question is about materials and components. *See pages 74–79. Spend about 10 minutes on this question.*

a)   Name a material that is renewable.   *(1 mark)*

b)   Name a material that is non-renewable.   *(1 mark)*

c)   Briefly explain how one of the materials you have studied is made from raw materials.   *(3 marks)*

d)   Describe one standard component you have used and explain why it was suitable. *(3 marks)*

2.   This question is about product analysis. *See pages 74–79, 88. Spend about 10 minutes on this question.*

The range of hats shown below was designed and manufactured for different target markets.

a)   Select *two* of these hats and describe the markets they were designed for.   *(2 marks)*

b)   Choose *one* hat and:

i) describe the properties of the material used in its construction   *(2 marks)*

ii) explain why it is suitable for the product to function well   *(2 marks)*

iii) suggest how it might have been manufactured.   *(2 marks)*

c)   Many products such as these hats carry brand logos. Explain why this might make the product more successful to the target market.   *(2 marks)*

d)   Products are tested to ensure they are fit for their purpose. Briefly describe a suitable test for *one* of the hats.   *(2 marks)*

3.   This question is about manufacturing. *See page 84. Spend about 10 minutes on this question.*

Each area of the Design and Technology department in your school has been asked to manufacture 300 identical, simple, low-cost gifts. These are to be given away to visitors during the opening of a new sailing club. They must be based on the motif shown on the right.

a)   Select a material you are familiar with and explain why it is suitable for this scale of production.   *(2 marks)*

b)   Using sketches and notes, explain the production process you would use to manufacture 300 of these gifts, using the material named above.

Marks will be awarded for:

*   accurate description of the process   *(4 marks)*

*   correct naming of tools and equipment   *(3 marks)*

*   clarity of drawings.   *(3 marks)*

4. **This question is about human factors.** *See pages 92–93. Spend about 25 minutes on this question.*

Consider the chair you are now sitting on.

a) Using sketches and notes, show what measurements of people might have been taken in order to design the chair for comfortable use.
(6 marks)

b) What word is used to describe the study of human measurement?
(2 marks)

c) Explain why a range of measurements is necessary when designing for people. *(3 marks)*

d) Not all products are designed to meet everyone's needs. There are many products designed to meet the special needs of minority groups.

i) Name such a product.
(1 mark)

ii) Explain what specific need the product is designed for.
(2 marks)

e) Many products are made adjustable to meet the needs of a wide range of differently-sized users.

i) Name a product which is adjustable in more than one way. *(1 mark)*

ii) Using notes and sketches, explain *two* ways the product meets the needs of a range of differently-sized users.
(10 marks)

5. **This question is about the care, storage and maintenance of products once purchased.** *See page 55. Spend about 10 minutes on this question.*

Many products require some form of care, storage or maintenance in order to increase their useful life.

a) Explain what *four* of these symbols below found on products, packaging or labelling tell the customer. *(4 marks)*

b) Many products require some form of coating to protect them from deterioration or to preserve their appearance.

i) Name a product which requires such a coating. *(1 mark)*

ii) Describe how the coating would be applied. *(3 marks)*

c) The products below require regular care and maintenance. With reference to one or more of the illustrations, identify *four* essential maintenance tasks that need to be undertaken regularly. *(4 marks)*

6. **This question is about packaging.** *See pages 104–106. Spend about 20 minutes on this question.*

a) Many products require some form of packaging. List *six* purposes of packaging. *(6 marks)*

b) Card nets are often used for packaging. Using notes and sketches, explain how these are cut out in commercial production. *(4 marks)*

c) Strong graphic design is an important feature of a lot of packaging.

i) Carefully draw the word 'Super' in a suitable style. Only *two* colours may be used.
(3 marks)

ii) Carefully draw the word 'Kids' in a style aimed at children. Only *two* colours may be used. *(3 marks)*

iii) Carefully draw the word 'Fast' in a style to suggest speed. Only *two* colours may be used.
(3 marks)

Project 3: Fun and games

Can you work as part of a team to help a company called 'GreatGames Ltd' to design and manufacture an exciting new board game, together with its packaging and promotional items?

The brief

'GreatGames Ltd' is a small manufacturing company that wants to expand its range of board-based games. It is looking for design ideas for games with strong visual themes and unusual 3D playing pieces. It is particularly interested in products that:

- are small and portable
- include mechanical or electronic components.

The company owns various CAD-CAM systems that it wants to make use of in the manufacture of the new games.

Where possible, the company would like to use recycled or reusable materials and components.

They are not interested in software programs for games consoles, but are willing to consider the use of PC multimedia content in conjunction with a product.

A brand new game concept is not expected: generally they will be based on existing traditional games, perhaps with modifications to the playing rules or using a new theme.

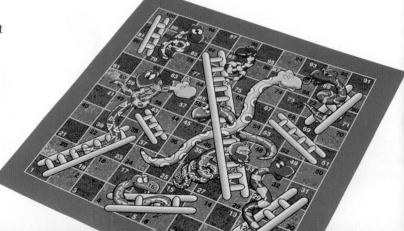

What you will need to do

- Conduct some general research into currently available board games, and analyse what makes them fun, popular and successful.
- Choose an appropriate theme and collect relevant visual material.
- Generate a series of ideas for new board-based games that meets the company's requirements.
- Decide on the final type of game and visual theme that you are going to develop.
- Construct a prototype version of the game that can be fully tested, evaluated and modified.
- Make a final version of the game using CAM wherever possible and appropriate.
- Set up a consumer trial to test your game out.

What you might also do

- Explain how a batch of 500 games would be manufactured.
- Use a DTP program to prepare a leaflet explaining the rules of the game.
- Use a graphics program to design the packaging for the game, and a promotional poster.

Depending on the time you have available, your teacher will discuss with you how many of the above tasks you will need to aim to complete.

IN YOUR PROJECT

'Fun and games' can be undertaken either as a team project or individually. It is suggested that team members collaborate on the initial research and development stages, and then perhaps take on responsibility for different end-products (e.g. the packaging, the rules, the playing pieces, etc.)

❛ The interactive Dr Who electronic board game! After the latest battle with the Daleks, the Tardis is malfunctioning badly and needs vital repair. You are the Doctor, racing through space and time, collecting six vital Tardis repair components that are needed to mend your time-travelling machine.

But beware, the Daleks are also roaming through space and are still trying to EXTERMINATE you!!! You must avoid them at all costs if you are to be successful in your quest. In this fast-paced action board game with interactive electronic Tardis and Dalek models, you must pit your wits against your arch-enemies. ❜

■ ACTIVITIES

1. Work together as a group to produce a list of as many different board games as possible. You should be able to come up with at least 50, but aim for 100! Do not include games that rely solely on cards (e.g. Snap), or computer games – there must be some real playing pieces, a board or mechanical or electronic product involved as well.

2. As a group, identify some visual themes that might be suitable for your board game. For example, horror, fashion, science-fiction, the characters from a book, film or TV programme, etc.

Making teamwork count

It's almost impossible to design something successful entirely by oneself. At some point, someone else's knowledge, experiences and abilities are going to be needed. A good team can achieve more than the sum of its parts, i.e. more than if everyone was working individually.

Team spirit

Working successfully with other people is not easy – it's a skill that needs to be practised.

Class acts...

Here are some key things that help to make a team successful:

- Everyone **agrees** on the general idea being developed.
- Members **listen** to what each other is saying and suggesting, and do not immediately dismiss new ideas, however crazy or impractical.
- Stronger members of the team **support** and **encourage** weaker members.
- Labour is **evenly divided** amongst the team members – no one is sitting around doing nothing.
- Everyone knows and respects the particular **specialisms** of each other.
- Things that are done have been **discussed** with, and agreed by, the team as a whole.
- Good **communication** between team members.

■ ACTIVITY

1. The Carrot Marketing Board wants to help promote carrots by producing a car sticker. It can contain text and drawn, coloured images. The size of the sticker is to be 127mm by 98mm (i.e. one third of a sheet of A4 paper).

   Work in a team of three or four to produce a sample design for a possible car sticker. For this exercise a computer may not be used to make the final design.

   This all needs to be completed in no longer than 15 minutes. It is suggested that:

   - five minutes are spent coming up with and deciding on the basic idea
   - five minutes are spent developing the idea and producing a rough version
   - five minutes are spent producing a final neat version.

Feedback

At the end of the 15 minutes, one person from each team will be required to give a brief account of their design solution to the rest of the class.

Lovely juicy carrots
Some fascinating facts

- The domestic carrot originated in Afghanistan in the tenth century, but these were commonly purple or yellow. The orange-coloured carrot emerged in the Netherlands in the fifteenth or sixteeenth century and became an emblem in the fight for Dutch independence.

- Carrots arrived in England during Elizabethan times. Some Elizabethans used their feathery stalks to decorate their hair, hats, dresses and coats.

- Carrots come in many sizes and shapes: round; cylindrical; fat; very small; long; or thin.

- Carrots were Britain's third favourite vegetable in 2005.

- The European Union defines carrots as a fruit as well as a vegetable.

- The legend that eating carrots helps you to see at night is believed to be 'misinformation' introduced during the Battle of Britain to cover up the secret invention of radar. However, a lack of Vitamin A, found in carrots, can result in impaired vision.

- The world's largest carrot was grown in 1998 and weighed 18.99lbs (8.164kg). The world's largest carrot statue is in Ohakune, New Zealand.

■ ACTIVITIES

2. How well did your team **work together**?
   How important and helpful was your **individual** contribution?

   In your team, spend 10 minutes discussing the following questions about how you worked together to design your car sticker for the Carrot Marketing Board.

   - Did anyone take on the role of guiding the overall planning and decision making?

   - Did everyone contribute to the process of coming up with initial ideas?

   - Did anyone suggest using the fascinating facts about carrots as a starting point for design ideas?

   - Was any one person better at coming up with the words that could be used?

   - Was any one person better at coming up with the illustrations that could be used?

   - How was the decision made as to which design to develop? How were any arguments and disagreements resolved?

   - In the development phase, how was the work shared out? Did everyone have something to contribute?

   - If one or more members of the team were not making a contribution, did anyone try to involve them in some way?

   - In the final production stage, did just one person make the sticker, or was the work shared? If so, how?

   - How was it decided who would present the design to the rest of the class?

3. Working in the same team, now spend 15 minutes designing some more marketing material based on your sticker design. For example:

   - Paper-hats and T-shirts.
   - Badges and key rings.
   - Food product recipes using carrots.

What makes games fun to play?

Studying some existing board games will help give you insights into the things that make them successful

Product analysis

Make a detailed study of a board game. You will need to have the game in front of you: do not work from memory.

Discuss your work with other members of your team, but produce your own analysis and evaluation sheet.

In your study, include comments on the following:

- The theme of the game.
- The market (i.e. age range) it is aimed at.
- How it has been manufactured and packaged.
- Whether or not it is well made.
- Whether or not the rules for playing are clear and easy to understand.

In particular, think about what makes it fun to play. For example:

- A strong visual theme in terms of graphics and playing pieces and a sense of role play.
- Very competitive, and not obvious who will win until the very end.
- Each 'play' lasts just the right amount of time.
- Involves skill and judgement, not just luck or chance.

As well as your own thoughts, take the opinions of the rest of your team into account, and also ask other friends or relatives. Discuss any differences of opinion in your study.

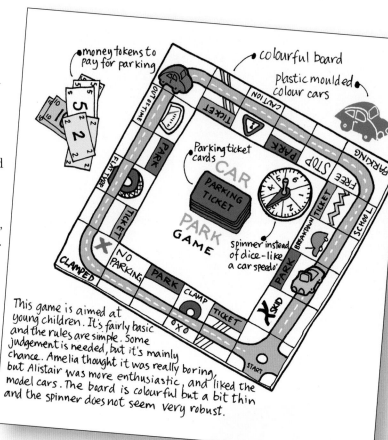

money tokens to pay for parking

colourful board

plastic moulded colour cars

Parking ticket cards

spinner instead of dice - like a car speedo'

This game is aimed at young children. It's fairly basic and the rules are simple. Some judgement is needed, but it's mainly chance. Amelia thought it was really boring, but Alistair was more enthusiastic, and liked the model cars. The board is colourful but a bit thin and the spinner does not seem very robust.

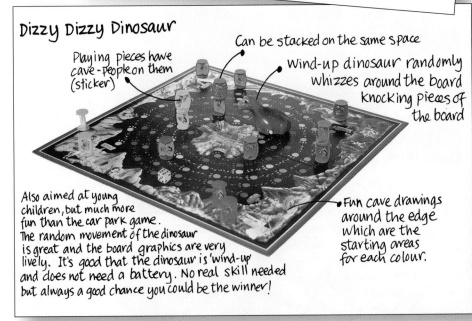

Dizzy Dizzy Dinosaur

Playing pieces have cave-people on them (sticker)

Can be stacked on the same space

Wind-up dinosaur randomly whizzes around the board knocking pieces of the board

Also aimed at young children, but much more fun than the car park game. The random movement of the dinosaur is great and the board graphics are very lively. It's good that the dinosaur is 'wind-up' and does not need a battery. No real skill needed but always a good chance you could be the winner!

Fun cave drawings around the edge which are the starting areas for each colour.

A brief history of card and board games

Did you know that card games were probably invented by the Chinese between the seventh and tenth centuries? They did not arrive in England until the mid-fifteenth century.

Draughts (or Checkers) has been in existence since 1400 BC.

An original version of Chess, using carved miniature horses, chariots, elephants and foot soldiers started in India and Persia about 4,000 years ago. Today's Chess game is about 2,000 years old. The current playing pieces

were designed in the mid-nineteenth century. The first computer Chess game program was written in 1951, but it was 1988 before a computer could beat a grandmaster.

Snakes and Ladders was one of the first popular board games, and invented in 1870.

The early history of Monopoly, the world's best-selling board game, is complicated. There is much confusion about earlier patented versions of the game, dating from 1904, and its subsequent

production development in the United States between 1925 and 1935.

The history of computer games is particularly interesting as it relates closely to the technological development of the computer.

The first space-themed game, called 'Spacewar!' was developed in 1962, but the first commercially-available game was based on table tennis and called 'Pong'. It was produced by Atari in 1973.

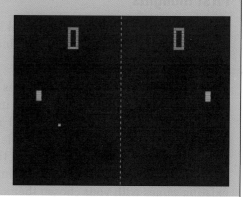

■ ACTIVITIES

1. Choose a traditional or modern game and see what further information you can discover about how and when it was designed. How much has it changed and developed since then?

2. Imagine you are writing a description of the game you analysed in activity 1 to be printed on the side of the box, or in a promotional leaflet. In no more than 75 words, see if you can explain how the game works and what makes it fun.

Theme sheet

Refer back to your list of possible themes from page 111. Each team member decides on a different possible theme for your game.

Work together to help collect pictures, images and samples of materials for each theme. Also contribute words, phrases and sayings that relate to the theme. If you have chosen a book, film or TV show, list the main characters, the location and any important objects.

Using the material collected by your team, assemble a reference collage sheet for your chosen theme, including some of the words, phrases and sayings.

Sustainability theme sheet

115

Developing great game ideas

Remember that 'GreatGames Ltd' is looking for new products that:
* have strong visual themes and unusual 3D playing pieces
* are small and portable
* include mechanical or electronic components.

They also want to be able to use their CAM equipment.
Make sure you keep this in mind while developing your ideas.

First thoughts

Start by quickly sketching some initial ideas. Remember, you are not designing an entirely original new game, but adapting an existing one in some way.

Make sure you have your lists of games from page 111 and your theme sheets from page 115 in front of you. Take an existing game (e.g. Chess) and adapt it to match one of your team's themes. For example, the board could become different regions of space and the playing pieces could be spaceships. Do this several times, matching existing games with your chosen themes.

Discuss the games with the rest of your team until some good ideas start to emerge.

Decision time

When you have finalised your possible games, each team member chooses an idea to develop further: it need not be the one you initially came up with yourself.

Getting down to details

Think about your chosen idea for a basic type of game and visual theme in more depth. How will it differ from the original: will it just look different or will the rules need to change too?

Making up the rules

Depending on what your idea is, consider the following:

* How many people can play?
* Will players need to throw dice?
* Can they decide what direction to move?
* Can other players be blocked?
* What 'chance' elements will there be?
* Are points scored on the way?
* How does a player finally win?

Changing the shape and form

Maybe the shape and form of the game will differ from the original? Consider and explore the possibilities of:

* using completely different materials and playing pieces
* making the board larger or smaller, or more three-dimensional
* including mechanical or electronic components
* producing a pocket-sized version.

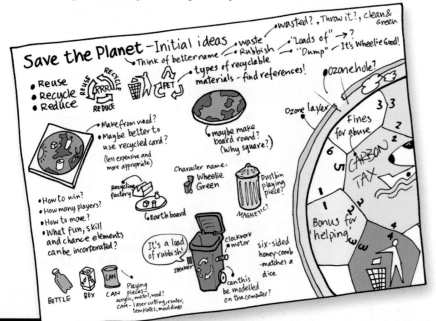

Early draughts?

Sketch out some possible layouts for the board. Use the images and colours from your theme sheet to make it as visually interesting and exciting as possible.

What does the team think?

What does the rest of your team think about the idea you have developed? How would they improve it if it were theirs? Listen carefully to what they say and incorporate their ideas if you think they improve the design.

The name of the game

Have you come up with a good name for your game yet? If not, look back to page 96 for ways of generating some possibilities.

Think about the original game you based your design on – maybe you can incorporate some of its name into yours? Or are there some words or phrases from your theme sheet that could be adapted? Remember, the name needs to sound fun and exciting to reflect the game.

Start thinking about the typography (lettering style) and logo that might be used to identify the game (see pages 97–98).

Using CAM

'GreatGames Ltd' needs to make full use of the CAM equipment they have in their workshops. Using CAM will enable them to achieve a high-quality finish, and to make batches of intricate product parts more quickly and efficiently.

As you develop your idea for a game, you will need to ensure it will make good use of their CAM facilities – even if you do not have access to such equipment when you make your final working prototype.

It is not necessary to use all of the equipment they have, but at least one machine must be used. Other non-CAM tools and equipment can be used as well.

Here is a list of the machines that 'GreatGames Ltd' has:

- A **vinyl cutter** – useful for making card nets for playing pieces and packaging.

- A **router** – for making 3D moulds for playing pieces or cutting out simple counters.

- A **laser-cutter** – could be used for cutting and engraving a wide range of sheet materials. it is especially effective with acrylic.

- An **embroidery machine** – could be used to embroider playing pieces on a fabric-based game for young children.

- An A3 colour **inkjet, laser and sublimation printer** – for graphics for the board, any playing cards, rules, packaging and promotional items.

- CNC lathe – for turning playing pieces.

Standard components

Products are usually made up of a number of smaller parts called components. Some of these can be made at a different time, and often in a different place, by another company.

What standard components could you use in your design?

Not made here

There are two main reasons why manufacturers may use pre-made standard components.

First, they may not be able to produce them themselves. For example, components that require specialist materials, skills and machinery are not readily available.

Second, although a manufacturer may be able to make them, it may be quicker and cheaper for a more specialised volume supplier to produce them.

For example:

- A textile manufacturer making casual jackets might buy in ready-made zips, buttons, Velcro and piping.
- A food manufacturer making pies might buy in ready-made fruit fillings or pre-shaped pastry.
- A product manufacturer making DVD players might buy in ready-made remote control units.

Choosing components

A designer needs to consider the following when choosing pre-made standard components for a product:

- The cost must be in the right price range for the product.
- Safety requirements must be adhered to.
- The size, weight and colour must match the style of the product.
- It must have similar performance characteristics to the rest of the product.

Advantages
of standard components

Manufacturers may use standard components:

- to save time
- because they do not have the necessary specific machinery or skilled workers
- so a wider range of products can be produced
- because complex production lines take up a lot of space and are expensive to set up.

Disadvantages
of standard components

- Time must be allowed for ordering and supply.
- Components are usually bought in bulk and have to be stored in the right conditions.
- They can be more expensive.
- The manufacturer is relying on another company that could let them down.

Making and testing prototypes

To find out how your game can be improved you need to make and test a prototype. Then you can go on to make the final version.

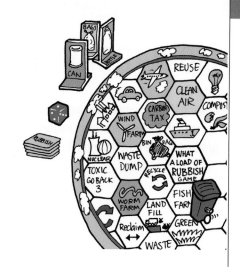

How well does it work?

Think carefully about some aspect of your design that you are not too sure about in terms of how it will work. It might be whether or not:

- the game will get going quickly enough
- it will become dull or repetitive at any particular stage
- it is too obvious early on who will win
- the playing pieces are the right size
- the layout and appearance of the board is successful.

What do you want to investigate?

Make a note of exactly what it is you want to test out. With this in mind, make a prototype that will enable you to discover what you need to know. Don't waste time on details that are not needed for your specific tests.

The prototype will probably be full size but made in different materials to the final version. For example you might:

- use paper instead of card
- use card instead of wood or plastic
- make things quickly and roughly by hand
- not include colour or artwork, unless it is essential for the test.

Fancy a game...?

Test your game out by getting your team to play it. Make careful notes about what happens. As well as the specific things you were testing, you might discover other things that need changing. Record the main comments made by your other team members. Explain whether or not you think their comments are justified.

Update your design to take account of the results of the tests of your prototypes. If you have not already done so, finalise the name of the design and its typography and any logo.

Planning and making the final version

Carefully plan how you are going to make a final version of your game:

- Are the materials you need easily available?
- What equipment will you need?
- How much time have you got?

Remember to use CAD-CAM as much as possible.

Aim to make your own game yourself as much as possible, but you can ask for technical help and advice from other members of your team. Remember to help and advise them in return.

How many do you want?

Making things in quantity requires a different process to making just one. Try this production process simulation to find out how to make a batch of products more quickly and more cheaply.

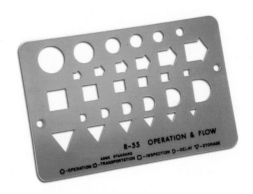

One at a time

When you made your final product you probably made a list of all the things you had to do. Then, very likely, you made each bit, one by one, and finally assembled them all together. This is called **one-off production**.

10 at a time

Suppose instead of making one product at a time you had to make 10 of them. You could make each one after the other, but it would be more sensible to work out how to make more than one at a time. This is called **batch production**. For example, you might:

- mark up and cut enough materials for all 10 products in one go
- make a template to make cutting an awkward shape quickly and accurately
- use a jig to help position a number of components together quickly.

This way, you would be able to make 10 products more quickly.

100 at a time

Think about what would happen if you had to make 100 products, but there were four of you working on them at the same time. This time, you would need to divide the tasks up between all four team members in the most efficient way.

One person might concentrate on preparing the materials, someone else on making a particular component, and the third person on assembling them together. The fourth person might be needed to check that everything was being made properly, to the required standard.

This way you would be able to make 100 products much more quickly than if one person was making each one from start to finish. This is the principle of a **production line**.

One of the main problems to avoid is to have one person waiting around for another to finish their job. It's also important to minimise the amount of movement of finished components around the workspace. Inspecting the quality at each stage is important too, to ensure all the finished products work as they should (see Quality counts on pages 122–123).

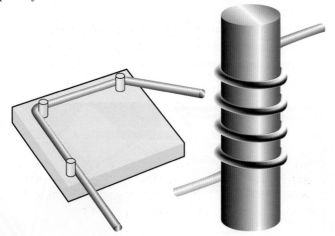

Production process diagrams

When planning a production line, a specific set of **symbols** is used to represent the different stages of making. In a complicated product, each stage of manufacture would have its own flow chart, with a master chart showing how the stages all flow together.

Slightly different symbols are used for writing a computer program or a control sequence.

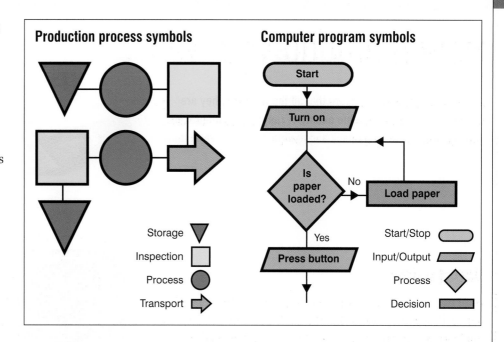

Production process symbols

Storage ▼
Inspection ☐
Process ●
Transport ➜

Computer program symbols

Start/Stop ⬭
Input/Output ▱
Process ◇
Decision ▭

■ ACTIVITIES

1. You need to make a batch of 20 'spinners' for a game.

Here is the basic sequence of production:

a) Mark and cut out a hexagonal shape from thick card.

b) Mark a series of numbers on to the card and add some colour.

c) Mark and cut out a short length of dowel rod.

d) Sharpen the end of the dowel rod.

e) Drill a small hole through the centre of the card, slightly smaller than the diameter of the dowel rod.

f) Pierce the dowel rod through the card to the middle of the rod.

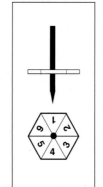

Work out appropriate sizes for the components according to the material you have available.

- Follow the production process to completely make a single spinner. Time how long it takes.
- In your team, design and make a number of jigs and templates that would enable you to make the different components in batches.
- Plan out how your team could work together to make a batch of 20 spinners in the shortest possible time. This involves working out:
  - who does what
  - where they do it
  - when components are checked
  - what to do if any components are not good enough
  - where to store components not being worked on
  - how to move components most efficiently between workers
  - what health and safety precautions are needed.

When you have finalised your assembly line, work together to make the batch of 20 spinners. Time how long it takes.

Compare your timing and the quality of your spinners with those done by other teams in your class.

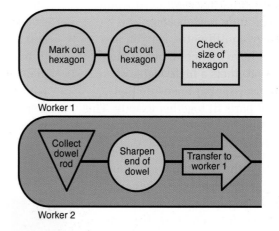

2. Assume you have access to CAD-CAM equipment to design and make 20 spinners. Explain how you would set up and use the equipment. Estimate how long it would take to make 20 spinners. How good do you think they would be?

Quality counts

Manufacturers need to ensure that all the products they are making are of an acceptable quality. A range of techniques has been developed to help check and maintain quality over a long production run.

Zero defects

The need for consistent quality in products is very important. A company needs to make sure it does not develop a reputation for manufacturing products that are unreliable. It also wants to minimise the costs of rejects and repairs. Gauges, jigs, templates and measuring systems (e.g. callipers, micrometers, electronic sensors, etc.) are used to check accuracy.

Using CAM systems greatly helps to achieve fewer defective products.

Working to tolerance

Have you ever tried making something exactly to size? It is unlikely that you succeeded: what you produced would almost certainly have been out by fractions of a millimetre. For you this might not have mattered, but in highly-complex products a high degree of accuracy is essential to ensure that important parts fit together exactly. The key question becomes: 'Exactly how accurate does it need to be?'

Tolerance is the amount of allowable variation in size from the original manufacturing specification. In some products, such as a high-performance engine, the tolerance level for each component needs to be very small. The more accurately a product is made, the better the quality in terms of performance and reliability.

In other products, such as a toy racing car, the tolerance can be greater. This allows for a wider range of acceptable sizes. For example, the wheel axle could be 70mm + 0.2 in diameter. '+' means 'plus or minus'. So a tolerance of 70mm + 0.2 diameter means that it can be anywhere within the range 69.98–70.02mm. If it were any larger or smaller, then the wheels would not fit.

Quality control

Quality control involves a series of methods used to check products as they are made. In most cases, it would take too long to inspect every item on a production line. For this reason, a sample is examined – one in every 100 or 1,000 perhaps, or once a minute or half-hour, depending on the product.

The results of the test are recorded and compared. If they become regularly unsatisfactory, then the production process is stopped. This feedback about the state of the production system helps identify problems before too many substandard products are produced.

Some quality control checks will be undertaken and recorded by production line workers. Computer-controlled systems are frequently used for machine processes. Computers can supply feedback quickly. They can process large amounts of information at high speed and keep logs of data over long periods of time that can be easily accessed and analysed.

Quality assurance

Quality assurance is the overall approach that a company takes to keep standards high in all aspects of its procedures, documentation, communications and product quality. Quality assurance provides a guarantee to consumers that all its products and services are of a certain standard.

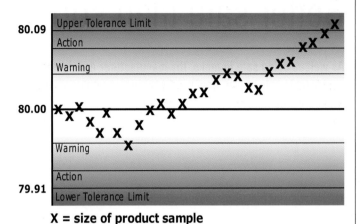

X = size of product sample

IN YOUR PROJECT

- How accurately made do the different parts of your design need to be?

- Which components need to fit together most accurately?

- At what stages of manufacture would you recommend that a sample of your product should be measured for accuracy?

- What inspection and measurement tests could be carried out?

- How often should the tests be done?

KEY POINTS

- Quality control systems help manufacturers reduce wastage and delay in production.

- They do this by predicting failure before it happens.

Better safe than sorry

As well as being safe to make, the manufacturer and designer need to ensure that a product is safe to use. New products must meet the requirements of the British Standards Institute.

Making it safe to use

All products and components are potentially dangerous in some way. For example, the possibility of sharp edges, small splinters, toxic finishes, exposed electrical parts, etc. all need to be eliminated in the design and manufacturing prototype stages. The general stability of a product is important too, to ensure it does not fall over.

Maintaining the standard

The **BSI** (**British Standards Institute**) provides detailed safety guidance and regulations for a wide variety of products. Their documents highlight the essential technical requirements for a product.

If a product is to be sold outside the UK, it will also need to conform to the required standards of each of the countries it is to be sold in. Similarly, products made outside the UK must conform to BSI standards if they are to be sold in the UK.

The letters CE show that the manufacturer claims that the product meets the essential safety requirements of the European Commission.

This symbol confirms that the product complies with the requirements of the BSI. The product needs to be independently tested.

Toy story

Safety is of particular importance in the design of children's toys and games. The requirements are covered in BS 5665. The standard has three main sections:

- Physical and mechanical – all the parts that can be touched.
- Flammability – how quickly it could catch fire.
- Choking hazards – parts that could be detached and swallowed.

A careful choice of materials and finishes is needed, and product stability is also very important.

Testing times

Before a product can be put on sale it has to be thoroughly tested to ensure it meets the requirements of the BSI standard. The test methods are laid down in the BSI specification, and are often carried out by independent organisations. Sometimes, special test-rigs are set up to quickly simulate use over a long period of time, so that the effects of wear and tear can be predicted.

After a product has passed its safety tests, the manufacturer can include the appropriate safety mark or symbol on the product and/or its packaging.

Which works best?

Consumer groups, such as the Consumers' Association, conduct their own tests on a wide range of products, including safety. As well as scientific performance testing, they also interview users.

BSI Education

Even skaters need Standards

All together now

'GreatGames Ltd' has now asked you to develop just one of the games designed by the members of your team for production. They have left the choice of which one up to you.

Which will it be, and how will 1,000 be made using the company's CAD-CAM equipment? The company has also asked you to develop proposed designs for the rules, packaging and promotion.

Difficult decisions

First, your team will need to agree on which of the games you have each designed and made will go forward into production. The decision might not be as easy as you think. For example, the game that is the most fun and exciting to play might:

- involve the use of a lot of expensive materials and components
- not be very suitable for production using CAD-CAM
- take a lot of time to make
- only appeal to a relatively small number of people.

You might need to agree to **compromise** by choosing a game that is not quite so good to play, but would be popular and can be made quickly and cheaply using CAM.

Note that you have been asked to prepare plans for the production of 1,000 games, not to actually make them!

We can work it out!

When you have decided which one of your game designs you will produce, you will need to agree on how to share the work out. Remember, you have to prepare the rules, packaging and promotion, as well as make the game. Read more about this on pages 126–127. Some tasks could be done by just one person, while other tasks might need to be undertaken by several people.

Who does what?

Discuss the particular **strengths** and **interests** you each have. Maybe one person:

- is very confident about how to plan for batch production using the CAM equipment
- is excellent at preparing artwork
- is very familiar with using graphic CAD packages to prepare layouts and print them out
- is good at having ideas for designing a promotional strategy
- enjoys writing instructions and advertising copy.

Again, some **compromise** might be needed to ensure that everyone has an equal amount of work to do.

Planning and recording

Ensure you know how long you have got to complete the task. Divide the time into a series of stages. Make sure you communicate with each other regularly to check everything is going to plan. If not, you may need to re-allocate some of the tasks to help others catch up.

As you work, keep a brief record of exactly what you've done, and how the others in your team are doing. Then, at the end of the project it will be clear as to exactly what your particular contributions have been.

All together now [continued]

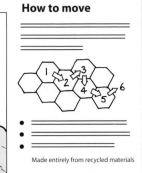

Production schedule

How would you advise 'GreatGames Ltd' to make the most of their CAM equipment in order to make a batch of 1,000 games? Prepare a short presentation that includes outline details of the production sequence, the use of special jigs and tools, quality control procedures and safety concerns (see pages 120–124).

Writing the rule book

Writing a series of clear, understandable instructions is not as easy as you might think. The order in which readers are presented with various items of information is very important, otherwise it just won't all make sense.

Start by making a short statement about the aim of the game and its theme. Try to explain what makes it fun and exciting, and how someone wins. Use short, simple sentences that just explain one thing each.

Think carefully about what **illustrations** to include, such as playing pieces, complex moves, etc.

Keep in mind that most people do not like having to read rules and instructions – they just want to get started!

Consider the **format**:

- What size will the paper be?
- Will it be portrait or landscape?
- Will it be folded?
- Will there be more than one piece of paper?

Finally, assemble the text and illustrations in a graphics or DTP program and print copies of the rules out.

 The rules will have to be included as part of the packaging of the game, so make sure you develop them alongside whoever is working on the packaging.

Packaging

- How and where will the different parts be housed in the box?
- What will the box be made from? How can size and waste be minimised?
- What will the box look like, e.g. shape, size, colours, images, typography?
- Are there any alternatives to a conventional six-sided box made from card?

Make sure you study the sections on packaging on pages 102–106.

The name of your team's game and its typography and logo should already have been finalised (see pages 97–99). If for any reason they haven't, or they need to be changed or improved, now is the time to do so.

Finally, make a one-off prototype version of the packaging. If possible, use a graphics and/or CAD-CAM program to print and cut it out.

 Remember that the packaging plays a very important role in promoting the product. Make sure you develop your design ideas alongside whoever is working on the marketing materials.

Marketing

However good your game might be, it's not going to be successful if no one knows about it. There are so many toys and games available in the shops, it's hard to get something new noticed. To help bring it to people's attention, some effective promotion is going to be needed.

- What sort of things could you give to a games store or local toy shop to help them display the product? Posters, leaflets and point-of-sale units are obvious possibilities.
- Can you think of any events or publicity stunts that would help attract attention and local newspaper coverage from the press?
- Who might suitably endorse the game? Think about its visual theme.

Electronic communications

'GreatGames Ltd' cannot afford TV advertising, but a short in-store video is a possibility. The company also has a website and a mailing list.

- How could 'GreatGames Ltd' encourage regular customers to spread the word around about the game using e-mail?
- Are there any promotional opportunities in the use of mobile phones?

Use DTP and graphics CAD programs to assemble a presentation folder containing your promotional ideas, including a poster. If you are making a video, you will need to use a digital camera and editing software. Transfer your video to a VHS tape, CD or DVD, or on to the internet.

 Make sure you work closely with the rest of the team to reflect the visual style and theme of the game and the packaging.

Final testing and evaluation

User trials

Working as a team, present your game to a group of children or teenagers. Invite them to open the box, find the rules and work out what they need to do. Show them the posters and other marketing materials. Listen carefully about what they are saying to each other about the game and make some notes. Aim to discover:

- how quickly and easily they worked out how to play the game
- whether or not they found it fun and exciting to play
- how long the game sustained their interest.

Presenting to the client

Write a short final evaluation report on the success of the game to put in a presentation folder for 'GreatGames Ltd'. The folder will need to convince the company that yours is the game they should make a serious financial investment in. It will need to include what they need to know about:

- the game itself
- what will make it successful
- how it will be made in quantity using CAM
- how it will be promoted.

How well did you work?

Finally, you will need to prepare your own personal evaluation of how well you worked on the project. How successfully did you:

- investigate a visual theme
- develop and test your ideas
- work as part of a team?

Project 4: Keeping in touch

Product design involves predicting the future: how can new and existing technologies be used to provide the products that people will need and want? Can you advise a company about what direction it should be moving in?

In Touch Telecommunications

CONFIDENTIAL MARKETING REPORT

Five years into the future, things will be pretty much the same. There will still be traffic congestion, hospital waiting lists and a hole in the ozone layer. Teleporting, time travel and 3D holographic TV are still a very long way off.

But one thing that will continue to change rapidly is the way people send and receive messages, information and entertainment. Instead of a multitude of mobile phones, portable playstations, DVD and mp3 players, everything is likely to come neatly together in just one hand-held device, linked into an almost invisible home network system.

Or will it? Although global communication technology companies are busy manufacturing and marketing the latest multi-purpose electronic gizmos, is this what people really want?

Keeping ahead of the market

Like many businesses, 'In Touch Telecommunications' works hard to keep ahead of the market. In the mid 1990s, the company was manufacturing innovative mini pagers that enabled teenagers to send simple text messages to each other. By the end of the century, they were developing the first generation of 3G mobile phones. In 2005, they moved into the design of Personal Digital Assistants (PDAs) and Smart phones with full internet access.

'In Touch Telecommunications' is now looking further ahead to the sorts of mobile personal electronic products that will become desirable over the next five years. To make a start, they commissioned a market research report. Now they have asked you to prepare drawings, prototypes, models and a presentation that shows what such devices might do and be like to use. They want you to make particular reference to how they will be carried or 'worn', and what 'accessories' might be appropriate.

1995   **2000**   **2005**   **The future**

Are these the electronic communication devices of the future?

This is a camera watch with a colour display that can take up to 36 pictures and store them until they can be transferred to a computer.

This solar pack, made from waterproof FineTex™, enables you to recharge your devices on the go.

The Gap Hoodio™ is a fleece jacket with a waterproof FM radio sewn into it, with a control panel on one sleeve, a power pack in one of the pockets, and detachable speakers in the hood.

Finger Phone™ is a small headset that you can wear like a ring on your finger. It can be used in a huge crowd to hold a conversation without having any trouble listening to the person at the other end. This is possible due to the bone conduction technology that the ring phone uses. If the user wishes to talk, they insert the ring into their ear.

What you will need to do

- Investigate how people use present-day multi-functional mobile phones and PDAs, including how they are **controlled**, **held**, **carried** and **protected**.
- Suggest the **functions** mobile communication devices of the future might have, and how they might be used in **specific situations**.
- Develop **imaginative** design ideas for the general appearance of such a device. Pay particular attention to how the device would be controlled, held, carried and protected.
- Suggest what **materials** such a product might be made from and how it might be **mass manufactured**.
- Produce an appropriate **presentation** of your proposals, including an appearance model of the device and its accessories and a series of explanatory display panels or electronic screens.

What you might also do

- Create a printed **instruction leaflet** and/or **sales brochure** for your product.
- Design and make the **packaging** and **point-of-sale** stand for your product.

Lumiloop™ is a modular system of program and display panels that can be chained together to form a reactive bracelet. Each display module features an LED matrix. The program modules hold different display programs and include varying sensors for the bracelet to respond to. A program module with an accelerometer interprets gestural motions of the wrist and generates illuminated patterns in response.

The iMan Cometh

In this article, Sheryl Garratt meets Jonathan Ive, currently one of the most successful and best-known product designers. He has transformed the way our personal electronic devices work and look.

See how far you can adopt similar approaches to your work in this project.

Jonathan Ive is from Chigwell in Essex. He dresses in a black T-shirt and trainers, and his official job title is Senior Vice-President in charge of industrial design at Apple Computers. If you met him, you'd find a down-to-earth, likeable man with a passion for music, and a habit of talking with his hands. What his unassuming manner wouldn't lead you to guess is that he is also the most influential product designer in the world.

In the past 15 years, Ive and his team have changed the way we think about computers, and they have completely changed the way we consume music with the iPod™. They have introduced so many new ideas, materials and techniques that you're bound to own something influenced by them, even if you've never owned a computer.

Tangerine dream

The son of a silversmith, as a child Ive was fascinated by how things worked. He studied industrial design at college in Gateshead. Afterwards, he co-founded Tangerine, a small independent design consultancy in London, where he worked on everything from toilets to televisions. But he found it frustrating because, by the time he was brought in, most decisions about how the product should work had been made. So when one of his clients, Apple, offered him a job, he felt it might offer him real influence. He would be working at the cutting edge of technology, designing products never seen before, with a company that seemed to share his high standards.

Apple: i is for internet

It didn't work out that way at first. When he joined the company, Apple was in trouble as a brand: sales were lacklustre, and many felt it had lost its way. Then Apple co-founder Steve Jobs returned to the company and asked them to start work on an easy-to-use, all-in-one home computer. Launched in 1998, the iMac was revolutionary. A gorgeous, retro-styled, colourful object that was a pleasure to use and to look at, it transformed the computer from a boring beige box to a stylish home accessory. It also turned around Apple's ailing fortunes.

Since then, Ive's team has consistently produced beautiful products that set new standards in computers and pioneer new manufacturing techniques that spread still further. What they specialise in is simplicity. Which, Ive points out, is harder than it looks. Take the iPod. There were digital music players before it. Trashy-looking plastic things with lots of buttons to operate and cables to connect, they appealed mainly to gadget-obsessed lads. Then came Apple's sleek, substantial version, which was controlled by a rather lovely click wheel on the front. As a feat of engineering it was breathtaking, but that's not what you noticed as a consumer. What you noticed was what it could do, which was store huge amounts of music and allow you to access and play it quickly and easily.

It starts with people

It was puzzling at first because no one had seen anything like the iPod, but as soon as you held it you intuitively knew how to use it: the fact that its front looked something like a stereo speaker even told you, on a subliminal level, what it was for. You quickly understood that here was a light device that fitted into the palm of your hand but could contain your entire CD collection. If you cared about music, you also knew you had to have one. Ive's products are often shockingly new in this way, but also strangely familiar. Summing up Ive's talent, eminent product designer Richard Seymour said this: 'It doesn't start with objects, it starts with people. And it starts with a very profound understanding of what makes us tick, how we think.'

Ive loves this aspect of his job: understanding how we relate to objects. To feel comfortable with it, we need to connect with it, to be able to relate it to products we've used before. He and his team spend a lot of time talking about such things: how to make a home computer feel less intimidating, how to send quiet messages telling the user what to touch without being too bossy.

Getting down to details

Every detail, from the materials to the packaging, is considered by Ive's team. If two components clip together, they will have thought about the pressure you have to use, the sound it will make. Every hinge or button is examined: Why is it needed? What might the problems be for the user? He shows me some screws on the inside of a computer case. They had to be custom-made, he says, so they wouldn't show. When you're manufacturing in bulk, it doesn't have to cost more to do this. But it takes time. You have to care. And you have to be working somewhere that allows you to care.

Innovation by design

When most companies talk about design, when they ask head-hunters to get them a Jonathan Ive, what they generally mean is they want someone to put the same old components into a shiny new package. They see him as adding a veneer of cool that sells products. But what Ive's team really does is something far more difficult. To truly innovate, you have to go back to the materials from which a product is made, and examine the ideas and assumptions that shape it. You have to ask what it is for, how people use it, and whether you can make that experience better. And then you have to pour enormous resources into seeing that through to production.

'People talk about how design is important, but that's such a partial truth,' he says. 'Firstly, it's good design that is important. The longer we do this, the more aware I become of how hard that is. But it doesn't matter how good the design group is if it's within a company that's not interested in doing something new and substantial.'

Into the future

Ive describes the pace of change in his industry as 'brutal'. Before a new product is in the shops, smaller, faster components are becoming available that will make it obsolete. I ask him for his predictions for the next 10 or 20 years, and he says he's usually too preoccupied by what he's making in the next two to see much further forward. 'What's clear is that stuff is getting faster and cheaper. For instance, you can now write music or do sophisticated video editing or digital image editing on a tiny laptop. We can all utilise power that a few years ago would have been only for professionals. I wish I'd had the computer I've got now when I was at college – man, what I could have done!'

He shows me the latest Apple professional computer, in a case made of one piece of aluminium with a hole punched to create handles. There is no plastic cladding, no extra pieces welded on, no visible screws or fixings. 'It's just a really honest piece of beautiful aluminium,' he says lovingly.

Eventually, I suspect there will come a time when he may want to step back from technology, to develop products that might still be used in 100 years' time. But not yet. 'We're always learning about different materials and processes. What we're seeing with some of this new work is the benefit of that cumulative experience.'

Jonathan Ive is 38 years old, and he's already made products that have changed the way we live. And the best thing is, you sense he's only just getting into his stride.

■ ACTIVITY

a) Read the article about Jonathan Ive. As you work though it, look out for details about:

- a team approach to design
- designing products that people want to use
- the importance of simplicity
- how every detail of the product is considered
- how computers are used in design
- how good design made Apple a highly profitable company
- Ive's predictions for the future.

b) Explain briefly in your own words what particular things make Ive's products so successful.

c) How might the way he designs influence the development of your communication device of the future?

Keeping in touch

Evolution by design

Most design is evolutionary. New products are rarely completely new, but an adaptation of an existing design. They might incorporate new features, a more aerodynamic shape perhaps, or use a number of different materials and components to produce a better-looking, better-working model that is cheaper to manufacture.

One good thing leads to another

It's very rare for a completely new product to come on to the market. Designs tend to develop over time, adding new features, safety devices, improved performance, or being sold more cheaply due to the use of more efficient materials and production methods.

Sometimes products **diverge** (i.e. move apart, or become increasingly different) – a new innovation can lead to a wide range of alternatives to meet a wide range of varying consumer needs. On other occasions products **converge** (i.e. come together and become increasingly similar).

During the twentieth century, communications technologies tended to diverge. At the start of the twenty-first century, they are rapidly converging.

Convergence by design

- **Cave paintings** have evolved into hand-drawn images on paper using ink, to block printing, laser-printing and graphic tablets.
- **Telephones** have evolved into mobile phones, to phones that can access the internet.
- **Radios** have evolved into portables, DAB and podcasting.
- **Televisions** have evolved from large items of furniture to colour sets, portables, flat screens, widescreens and mp4 players.
- **Wax cylinders** have evolved into audio and videotape cassettes, mini DV, DVDs and flash memory.
- **Cameras** have evolved into Box Brownies, 35mm pocket cameras and digital movie cameras.
- **Computers** have evolved from being large mainframes, to desktops and hand-held devices connected to the internet.
- **Calculators** have evolved from the simple abacus into mechanical adding machines, electronic calculators and PC programs.
- **Clocks** have evolved into analogue and digital watches and stop-watches.

Until recently, these were all different products. They are now converging into just one or two products that can be carried around in your pocket.

■ ACTIVITY

1. Identify an electronic communication device first introduced in the twentieth century: for example a telephone, television or computer. Alternatively, choose another twentieth-century product that uses electricity in some way, e.g. a kettle, power-drill, sewing machine, etc.

   At what points has the design development of the product been the result of divergence or convergence? Produce an illustrated chart or display panel that shows its design evolution.

The four phases of a product

When a new product is first launched, sales are likely to be slow. Profits from sales are unlikely to provide a return on the development, mass manufacture and initial promotion costs.

In the second phase, demand picks up as the product becomes known and accepted. Everyone starts to want one, and sales repay the initial investment and start to show a healthy profit.

In the third phase, the product will be well established and selling well with minimum promotion. However, it will also now have to compete with newer models being produced by other companies. These may well be cheaper or have a better performance specification.

In the final phase, sales will drop off. This can be rapid as rival products, now in their second and third phases, begin to dominate the market. The cost of production may exceed the potential profit, so the model is withdrawn. Sometimes manufacturers re-launch products with minor modifications to help extend the sales period.

Products are not intended to last for ever!

Market pull / Technology push

New products tend to evolve in one of two ways.

Sometimes consumers begin to demand particular types of products, or the features they want them to have – for example, more energy-efficiency, recyclability, fair trade, organic, etc. Manufacturers develop new products to meet this demand. This is called 'market pull'.

Meanwhile, new developments in technology, such as digitally available music, easy care fabrics, smart materials, etc., provide manufacturers with the possibilities of new products. These are presented to consumers as being desirable. This is known as 'technology push'.

Organic food products are demanded by consumers, and supermarkets are keen to respond.

People do not request a particular new fragrance. It is scientifically developed and pushed at them by promotion.

■ ACTIVITY

2. Identify an electronic communication device available on the market today or during the past five years. Using the internet and other sources of information, find out what you can about the history of the design development of the device.

   a) When was it first introduced in its present form?
   b) Was it the result of a technology push or a market pull?
   c) What phase of its availability in the market is it in?

Communication matters

- How and why do people need and want to communicate with each other?
- Are people satisfied with the devices currently available?

To find the answers to these questions, you need to compare and contrast some existing electronic communication devices.

Any time, anywhere…

Personal Digital Assistants, mp3 players, Smart phones, etc. all share one basic purpose – they exist to enable the user either to communicate with someone else and/or to receive information from others. The content that is communicated and received can be spoken words, sounds, still or moving images or numbers. Providing the sender and receiver each have access to the appropriate device and connection, the content can be transmitted anywhere across the globe in seconds.

What, when and how much?

The media we decide to use to communicate and access information is influenced by a variety of things:

- The nature of the content (i.e. visual, aural, numerical).
- The ability of the receiver to access, retain and understand it.
- The time it will take to prepare it.
- The cost of sending it.

Previously, there were only a limited number of ways of sending a personal message over a distance – mainly writing a letter, or making a telephone call. Now there are many ways. New electronic communication devices need to make it easy for a user to create appropriate and effective messages using different media, and for them to be able to receive, understand, organise and respond to content sent by others.

■ ACTIVITY

Obtain a number of electronic communication devices from different periods, e.g. from the pre-1980s, 1980s, 1990s and the start of the twenty-first century. They can be all of one type, e.g. types of telephone, or a mixture of different communication devices.

Make a study of them to compare and contrast their success or failure. Focus on how well the **displays** and **controls** have been designed to make them as easy to use as possible. How well do they meet the many **differing needs** people often have?

Your study should be based as far as possible on the comments and opinions made by other people, not yourself. These people may be others in your class or group but, ideally, it should include some older relatives and friends of the family.

You will need to ask a series of factual and opinion-based questions, such as:

- how often do you use this device
- where is it kept when not in use
- what accessories for it do you have and use
- do you like using it
- is it easy and comfortable to use
- what particular features do you like
- what particular things about it do you find frustrating or annoying?

If the object is no longer in use, these questions will need to be in the past tense, e.g. How often was this device used?

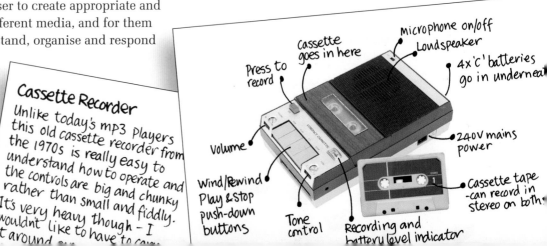

Cassette Recorder
Unlike today's mp3 Players this old cassette recorder from the 1970s is really easy to understand how to operate and the controls are big and chunky rather than small and fiddly. Its very heavy though – I wouldn't like to have to carry it around...

Press to record
Cassette goes in here
microphone on/off
Loudspeaker
4x 'c' batteries go in underneath
Volume
240V mains power
Wind/Rewind Play & stop push-down buttons
Tone control
Recording and battery level indicator
Cassette tape -can record in stereo on both

Anything you can do, I can do better

'My new mobile phone does just about anything, but it is more annoying than the old one. The new fangled colour screen won't show me the time, or whether there are any missed calls or messages without actively pressing a button to make it light up again. Also, the camera button keeps activating in my pocket so I end up with loads of pictures of blackness, while making people suspicious of me when I keep making camera motor-drive sounds – that strangely I don't seem to hear until someone says something!

Yes, of course I'd like an all-in-one camera / phone / TV / remote / watch / compass / gps / organiser / computer / walkman-radio with flip/roll out flexible screens and laser eye-coordinated controls. But, until then, just give me a phone that can make calls and lets me know when I've got a message or missed call by just glancing at it once in a while when I'm near it. I don't want to be permanently attached to it and have to evolve a clenched mobile phone holding grip and longer uni-dexterous thumbs to text with all the time.'

Displays and controls

- How is it switched on and off?
- What other switches, buttons, keys, etc. does it have? How well do they work? Are they easy to understand and operate, or small and fiddly?
- What audio signals (i.e. sounds) does it produce? Can these be easily heard, or can they be obtrusive?
- Do cables need to be attached to it? Is it easy to work out which ones go where? Do they get in the way, or become easily tangled or bent?

 Before you consider these questions make sure you read 'Human factors: displays and controls' on pages 136–137.

Differing needs and desires

Remember it's other people's thoughts and comments you want, not just your own. If different people give different responses to your questions, suggest why that might be. For example, they might:

- be more familiar with that type of device, or less experienced in using one
- not have much need or desire for such a device
- prefer older devices that were easier to use, and don't need the extra functions of the newer model.

Finally, if you have not already done so, ask them about what sort of communication devices they would like to have in the future. Are they looking for something that does everything, or just a couple of products that are easy to use? Can you see any ways in which they might find a new device useful, that they have not thought of?

You will need to think carefully about the best way to present your study. Aim to make it as visual as possible, using illustrations, charts and diagrams.

Old fashioned phone

This is my Mum's old phone that she kept from when she was little. In those days all phones were like this. The dial mechanism has a really nice feel to it and makes a great noise as it winds back. However, it takes ages, and would have been really annoying if the number was engaged and you had to re-dial each time. But my Mum says in those days you could only afford to make local calls, and they just had four numbers, so it wasn't that bad.

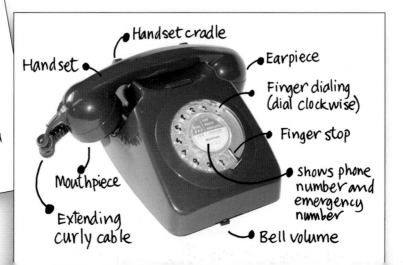

- Handset cradle
- Handset
- Earpiece
- Finger dialing (dial clockwise)
- Finger stop
- shows phone number and emergency number
- Mouthpiece
- Extending curly cable
- Bell volume

Human factors: displays and controls

Effective displays and controls are essential if people are going to make electronic products work properly. Displays provide the operator with important information about the state of the machine. Control devices can enable them to make the necessary adjustments.

A very important part of the job of the designer is to ensure that products are easy to understand and use.

A simple on/off switch can take many different forms.

Press the button

Most mechanical and electronic products work by means of a series of events happening in sequence. The movement of a number of mechanical devices and the flow of current in electronic circuits need to be controlled in a variety of ways – switching them on or off, making them work faster or slower, for example.

What a display or control looks and feels like, how it is operated and where it is placed can all make a great deal of difference. Often it's not so much the basic displays and controls used everyday that can cause problems, but the ones that are used infrequently and sometimes have to be set in a hurry or in an emergency.

At the interface

The displays and controls of a product are often called the **interface** – the point at which the person and the machine interact.

- **Displays** (such as lights, graphic symbols next to switches and dials, and sometimes synthetic voices) pass information to people about the current state of a machine or device. Is it on? How hot is it? What setting is it on?
- **Controls** (such as switches, dials and touch-sensitive buttons) enable people to alter the state of the machine, e.g. switching it off, cooling it down, changing the setting.

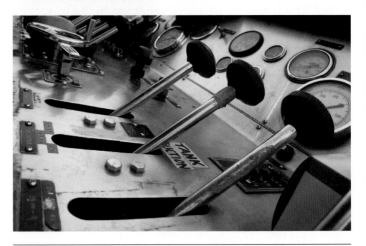

This kitchen timer combines a control mechanism with a visual display.

■ ACTIVITIES

1. Study the photographs above. Which are displays and which are controls? Which, if any, are both? How have graphic symbols been used to help display the state of the controls?

2. What devices can you think of that you have tried to operate, but were unable to get to do exactly what you wanted? It's not that it doesn't work, just that you don't know how to work it. Illustrate your answer with sketches and notes to explain what the problem was.

Control panel layout

Different displays and controls are used at different times while the product is in use.

Each one can be identified as being either:

- **on-line**, i.e. those which are used while the device is in use to monitor and control the system
- **off-line**, i.e. those which need to be used immediately before the product is used, or when it is due to be finished with
- **danger**, i.e. those which warn that something is wrong
- **maintenance**, i.e. those which are used periodically to ensure everything is working properly.

These displays and controls are often grouped together. For example, all the on-line displays and controls might be placed where they are easiest to get at. Sometimes the first off-line control, used to switch the device on, is on the far left-hand side. Warning displays are often placed with the on-line controls so that they will not be missed while the device is being used.

On-line controls.

The off-line controls used to set up a machine or device might be placed together at the side.

Danger signals must be easy to see.

IN YOUR PROJECT

Think carefully about how someone will need to use the displays and controls you are designing:

- Will they be easy to turn on and off, and to adjust?
- Would visual or aural displays be most effective?
- Where will be the best place for them to be placed?

■ ACTIVITY

3. Analyse the layout and arrangement of the displays and controls for your mobile phone, or a similar electronic device (e.g. games console, PDA, etc.)

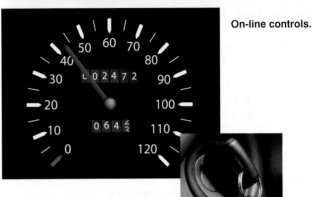

Imagine…

What are your initial thoughts for new communication products? Who might they be used by, and in what circumstances?

First thoughts

As you made your study of communication devices, you probably had some initial ideas for new products. Maybe they were for:

- a multi-functional device
- a device that could be used in a specific place or circumstance
- something that could be worn or carried in an unusual way
- something that would be easy to use
- possible accessories.

■ ACTIVITIES

1. Make some quick sketches of any ideas that you've had so far. Show them to other members of your group and get their opinions on them.

    - Which do they think they are the best ideas?
    - How would they take them further?
    - What problems need to be solved?

2. Studying some other products that solve similar problems is likely to give you some good ideas for your design. Look at each of the photographs on this page and explain:

    - how they can be used for more than one purpose
    - how different parts are linked or connected together
    - if they can be used anywhere, or only in a specific place
    - whether they hide or show what they do and how they work
    - where any displays and controls are located
    - how they can be easily carried or worn
    - how they look and feel – what emotional responses you have to them.

Ideally, find some objects of your own to examine.

Produce some further sketches to show how you might apply some of the solutions used in other products to your problem. Show your development work to your group again for their comments and suggestions.

Target markets

Who are the intended users of your product?
They might be:

- doctors on call, visiting patients
- hill walkers out on a day trip
- 10 year olds on a school trip
- college students on a study visit
- a teenager who belongs to a local band
- an estate agent estimating the value of a house.

Remember that what you are designing is not for you to use yourself – you must consider the needs of your target market. You may need to do some further research to find out more about their specific needs and desires.

' My device is intended for an elderly person to use. It must be very simple to operate, and look friendly rather than technological.

It will include an alarm system to enable them to call for help and satellite navigation to enable an emergency service to locate them immediately. Although less important, it could also be used as an alarm to remind them to take their medication: this might be an accessory. '

■ ACTIVITY

3. Decide which ideas to develop further. Consider and write a clear statement about each of the following:

   - What sort of person (i.e. target market) is most likely to use it, and what are their particular needs?
   - Where and how is the intended user most likely to use it?
   - What functions and features will the device need to have, and why?
   - Which functions and features are the most and the least important?
   - What sort of look and feel is most likely to appeal to the target market?

Going soft

As soon as you've got some good ideas on paper, or maybe even if you're getting stuck, try making a series of simple models in card or other basic modelling materials.

Start working out:

- its shape, form, size and scale
- the possible positions of displays and controls
- where and how the device will be kept when not in use
- what accessories might be useful.

Make sure you take some photographs of the models you make.

 See pages 60–61 in the Project guide.

Working smarter

A range of exciting new materials has been developed over the past 20 years. There are two main types of new materials – composite materials and smart materials.

IFM's Electronic Plaid™ is a beautiful, soft textile artwork that changes colour.

Composite materials

Traditional materials all have particular properties that are fixed. Now, new materials can be made to have a particular property that a designer wants. This means that the designer can specify the properties that are required for the job, rather than be restricted by the existing properties of standard metals, woods, plastics, paper and fibres.

These new materials are usually **composites**. This means they are created from different mixtures of other materials and chemicals that combine together to give the qualities that are required.

For example, conductive foam is used to protect integrated circuits when they are being transported. Their legs are inserted into the block of conductive foam. 'Normal' foam

would protect the circuits from knocks and bumps during transportation, however, conductive foam has been designed to have properties that stop the build up of static electricity – as well as protecting the integrated circuits from damage.

Modern manufacturing technologies are creating exciting new finishes for paper and boards. For example, the finish on inkjet papers can provide a very high quality for photographic reproduction.

Great advances are being made in hi-tech fabrics. The most recent technologies can combine warmth with lightness, breathability and water repellence in one fabric.

Microfibres

Mixing polymers to produce fibres that are even finer than silk produces microfibres. They can be woven so closely together they can prevent water from permeating, but also have high moisture transference properties so they still allow perspiration to escape. Microfibres are used for items such as lens-cloths and weatherproof fabrics for various outdoor activities. Microfibre cloth hangs well and is extremely soft, lightweight, durable and comfortable to wear.

Smart materials

The properties of smart materials change in response to a change in their environment, for example, temperature, movement and/or light. Some familiar examples of these materials are the plastics used for sunglasses that get darker when used in bright sunlight and ceramics and plastics that change colour when heated.

Many applications of smart materials include electrical components embedded within the material. These can be connected to computer systems.

Smart materials are often created to perform specific tasks in response to particular situations. They are also known as active, adaptive, intelligent or sensing materials.

Electronic paper

Special new inks can change colour when a small amount of heat is applied. Paper can be printed on with special inks that respond to an electric charge.

IN YOUR PROJECT

Describe the properties you would need for the composite or smart materials to be used in your product. Explain how they would improve the performance characteristics of your design.

This square of fabric acts as a computer keyboard. It can be folded, crumpled and even washed. As such, it could become part of a garment.

Hi-tech fabrics

Smart materials are able to change their properties and characteristics depending on their environment. For example **thermocratic** printing inks, which react to heat and change colour depending on the temperature can be used on fabrics. Fragrances can be added to fabrics. This is called **microencapsulation**.

Garments may be able to **display** text and images directly on the fabric using materials composed of a matrix of microscopic electronic beads. These pick up radio transmissions that control the colour and brightness of each bead.

Garments and textile products that incorporate computer technology may become more common. For example, jackets worn by mountain climbers may be able to send signals out that the wearer is in trouble. Someone old or unwell may receive signals that their temperature is dropping or they need to take some medication.

Motion control gels and gases

Motion control gels and gases, for example 'smart grease', are used to regulate the movement of components that are in contact with each other. Some examples are sliding microscope barrels, variable resistors and slow spring-return CD drawers.

Shape Memory Alloys

Shape Memory Alloys (SMAs) change shape and volume in response to certain conditions, for example temperature change. They can be 'trained' to reverse their transformation, returning to their original shape and size. These are often used in wire form to make springs and interwoven with fibres to make fabrics that change shape and size according to temperature. Bendable spectacle frames are made from SMAs.

Making it by computer

An increasing number of production processes can now be carried out by machines that are controlled by computers. Automated manufacturing is safer, quicker, more reliable and, in the long run, cheaper.

Computer-aided manufacture

Computer-aided manufacture (CAM) is a term used to describe the process whereby parts of a product are manufactured by equipment that is controlled by a computer.

One of the restrictions of batch production is that after a relatively small number of products have been made, a machine has to be re-set to the requirements of a different product.

The main advantage of CAM is that the new instructions are stored electronically and can be downloaded and programmed into the machine very quickly. This also facilitates making small changes to the design to suit changes in the market or to produce specialised short-run products for individual clients.

Where CAM is used to replace a manual operation, greater productivity is possible, because the machine can work continuously. There is also a greater consistency of quality, and fewer faulty goods. CAM systems can also work with materials and chemicals that might be harmful to human operators.

In many cases, the manufacture of a complete product is a combination of CAM and hand operations.

Computer numerical control

Computer numerically controlled (CNC) machine tools can be independently programmed, but can also exchange data with other computers. Therefore, they become part of a complex automated production system. This is particularly effective where a number of smaller manufacturers specialise in the making of a component that contributes to a whole product.

Remote manufacturing

Computer-generated data about a final design can be sent almost anywhere in the world in a few seconds. The data is fed directly from the computer into manufacturing equipment to make the product.

Computer integrated manufacture

Computer-aided design (CAD) and CAM systems contribute to the development of computer integrated manufacture (CIM). This is a totally automated production process with every aspect of manufacture controlled by computer.

Powerful CAD systems can be linked into a CIM system. This allows the entire design development, production schedule and manufacturing operation to be undertaken by a single system.

Manufacturing companies that have adopted such systems have been able to make dramatic reductions in their prices, and increase their quality and reliability.

Automated and autonomous guided vehicles can be used to transport components, tools and materials to the appropriate machine.

Automated computer systems are used widely in the textiles industry. For example, CIM systems enable colour separations for printed fabrics to be saved digitally. Designs for woven fabrics can be sent to a computer-operated loom.

Case Study douglasarkwright

From design to manufacture: 'His and Hers' USB flash drive pens

A design company, 'douglasarkwright', was commissioned to combine the designs and the technical specification for 'His and Hers' USB flash drive pens into viable products.

The first stage was to accurately measure the USB circuit board that would be used for the casing design. The circuit board has holes for pins to sit and hold it in place. Sketches and hand models were drawn up to get an idea of size, form and proportion for the pen and to convey the design intent. These sketches were developed into a 'soft' appearance model. The models were made using Styrofoam and a denser material called Greyfoam which can be cut, shaped and finished easily by hand.

Rendered images of the casing and components were constantly assessed to ensure that the model met the aesthetic, as well as the technical, requirements. Virtual assemblies were used to check that all parts fitted correctly and avoided any 'collisions'. Rapid prototyped parts were created using a 3D printer that slices the CAD model into 0.07mm layers and uses an acrylic paste that is cured through the use of UV light. These were used to check that all the electronic components and housings fitted correctly and that the tolerances used were correct.

Fully photorealistic images were produced. Colours, textures and graphic decals were assigned to the surfaces of the digital model. These images, along with the finished rapid prototype model, were used to advertise and promote the design on the internet at douglasarkwright.com.

Finally, the 3D digital data was sent via email directly to a toolmaker in the UK where the injection moulds were produced.

douglas ⅄arkwright

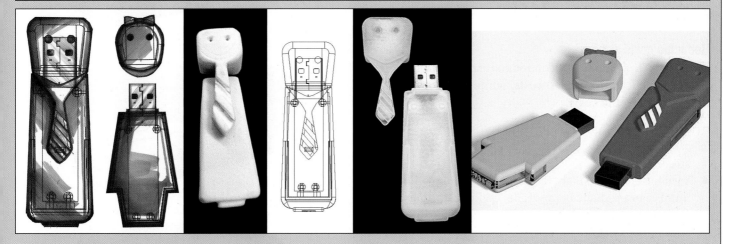

Advantages of CAD-CAM

As a result of using CAD-CAM:

- the process of design development is speeded up and, as a result, a greater number of designs can be produced
- there is an improvement in the quality of design, because the computer can more accurately simulate and produce information about how a design will behave in different operating conditions
- any changes made can be quickly communicated throughout the team working on the design
- design information generated on the system can be stored on electronic media and quickly and easily retrieved at a later date
- production is quicker, continuous and less labour-intensive
- a greater consistency of quality is produced
- machines can be reprogrammed very quickly to perform different production tasks.

Disadvantages of CAD-CAM

- Computer technology is expensive to buy and maintain.
- Extensive and costly training is often needed.
- Data can be lost if not properly organised and backed-up.

Design for manufacture

There are many different things that need to be taken into account when designing a product suitable for manufacture on a large scale. These include planning for the available production equipment and materials, the expected life of the product and the most cost-effective method of manufacture.

How will your product need to change to make it more suitable for batch or mass production?

Manufacturing constraints

Designers often have to design within considerable constraints, such as costs, the availability of resources, the requirements for large volume production and quality control.

Reducing costs

Frequently, the emphasis is on reducing production costs. Some elements will be **fixed costs**, while others are known as **variable costs**.

- Fixed costs are those incurred in setting up an assembly line, such as machines, tools and factory space.
- Variable costs are likely to change according to the number of products being made, and cover things such as raw materials, energy, staff wages, insurance, maintenance, etc.

The costs of storage, packaging, distribution and selling all need to be considered too. VAT is another element which adds to the final selling price of the product.

The actual manufacturing cost of a product in terms of its materials and labour will vary according to the particular item. Often, it only represents some 5–10% of the final selling price.

The profit made by the manufacturer, or investment organisation, will also vary depending on the product. Typically it ranges from just a few per cent for high-volume, rapid turnover goods, to up to 50% or more for exclusive, high-quality, hand-finished items.

"We review and evolve the design of our products frequently. Our aim is to produce products that are easier and more cost-effective to manufacture in large numbers, use fewer parts, and keep up with modern trends."

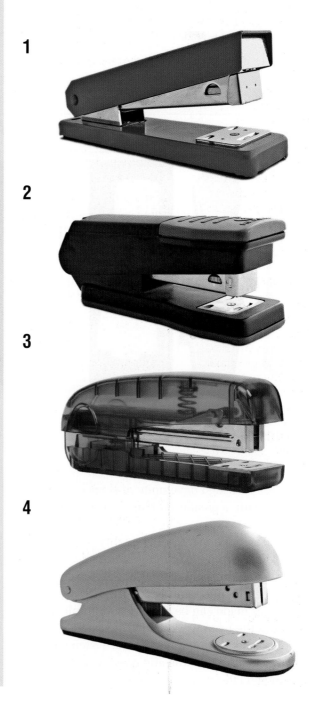

1

2

3

4

IN YOUR PROJECT

How would you set about designing your product so that it could be made:

- cheaper   • quicker
- better   • safer
- more desirable?

1 "Our original all metal model, made in the 1970s."

2 "Our mid-1980 model had an improved rubber base to reduce surface scratches, and included a plastic moulding for greater comfort in use."

3 "Towards the end of the 1990s, transparent injection-moulded plastics were being widely used. This model significantly reduced the amount of metal used, and the number of assembly operations made it much lighter, and cheaper to make. Batches of different colours could be easily set up to respond to consumer preferences. This model was designed using CAD."

4 "Our latest model has a more sophisticated appearance in line with current trends for black and silver, and for more organic shapes and forms. This design looks heavier and more durable, but the casing is still injection moulded in plastic. It was designed and made using our fully automated CAD-CAM system."

■ ACTIVITY

Identify a well-established product that has changed over time to make it more suitable for manufacture – for example, a hair-dryer, electric drill, jacket, etc. Identify the main changes that have been made to make it quicker and cheaper to make in large quantities.

Design for quantity

Products need to be designed to be easy to make. Some items may prove to be very difficult to make in quantity, however. This might be because of:

- their shape
- the way the components are arranged
- the materials required.

Different manufacturing processes and materials can be used according to the numbers to be produced.

The rate of production is an important factor too. Producing 10,000 units by the end of the week in order to satisfy demand will need to be approached in a different way to making the same number over a 12-month period. There are also important implications if production needs to be organised in batches, for example 5,000 this month and another 10,000 next month.

Design for maintenance

Designers need to consider how often a new product will need to be maintained during its usage, and take this into account while developing ideas.

They will also have to think about how easy it needs to be to undertake the maintenance work. If a component needs cleaning, adjusting or replacing often by the user (e.g. replacing a battery), it must be quick and easy to do. Other maintenance might need to be done by trained specialists, however, and providing easy access might result in damage if the user tries to do it by themselves.

Ideally, a product should be maintenance free, but this is likely to involve the use of more and higher-quality components and tolerances in manufacture. This will inevitably increase the cost.

Design for life expectancy

Customers expect a certain minimum time that the product will work for, which extends beyond the guarantee. This time will vary according to the product.

A product that fails before the end of the guarantee period is going to be very costly to the manufacturer to repair. If a product needs expensive repair soon after the guarantee expires, a customer is unlikely to make a repeat purchase of the same brand, and the brand might develop a reputation for being unreliable.

However, if a product works successfully for many years, consumers will not need to buy replacements so often, and demand will fall. The number of products made will drop and, as a result, the price will rise.

Many products contain components that are likely to fail after a number of years, and which would be very expensive to repair or replace. This is known as **planned obsolescence**.

Design limitations

Designers rarely have a free choice of materials, components and production processes, and often have to work with what a manufacturer already has.

Further limitations may be imposed by things like the maximum size a machine can mould, an existing stock of ready-made electronic components and the existing skills of the workforce.

Some products are created with the prime intention of utilising materials and equipment that are being under-used during the decline of a particular product.

Finding fault

Much work goes into improving the design of a product. Sometimes a fault may not be so much with the original design, but in the quality of manufacture.

Methods of manufacture may also be analysed to determine potential modifications to reduce environmental pollution and improve product sustainability.

Sometimes products fail to sell. Perhaps it quickly became known that it didn't work well or was unreliable. Maybe not enough money was invested in promotion, with the result that not enough people knew it was available. Possibly manufacturing costs proved to be much higher than expected, with the result that no profit was made.

Getting down to details

As well as sketching your ideas, try making a series of simple models
in card or other simple modelling materials to work out:

- shapes, forms, sizes and scales
- the possible positions of displays and controls
- where and how the product will be kept when not in use
- what accessories might be useful.

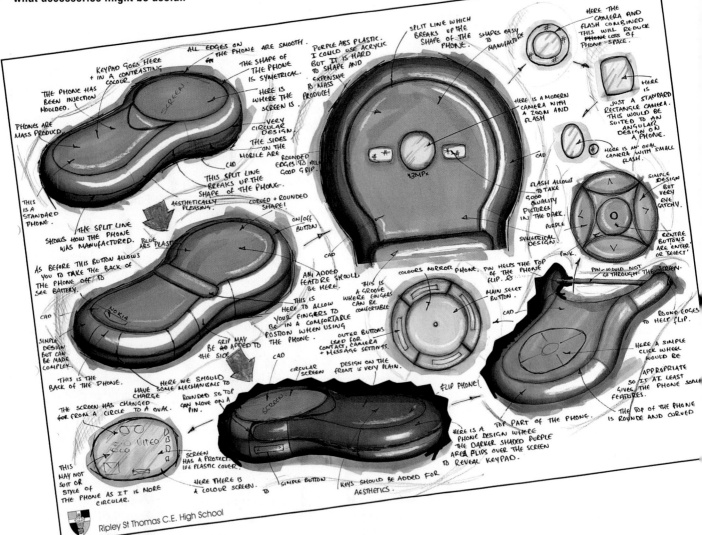

Ripley St Thomas C.E. High School

Make sure you take plenty of photographs of the models you make.

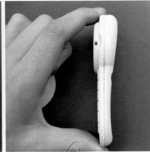

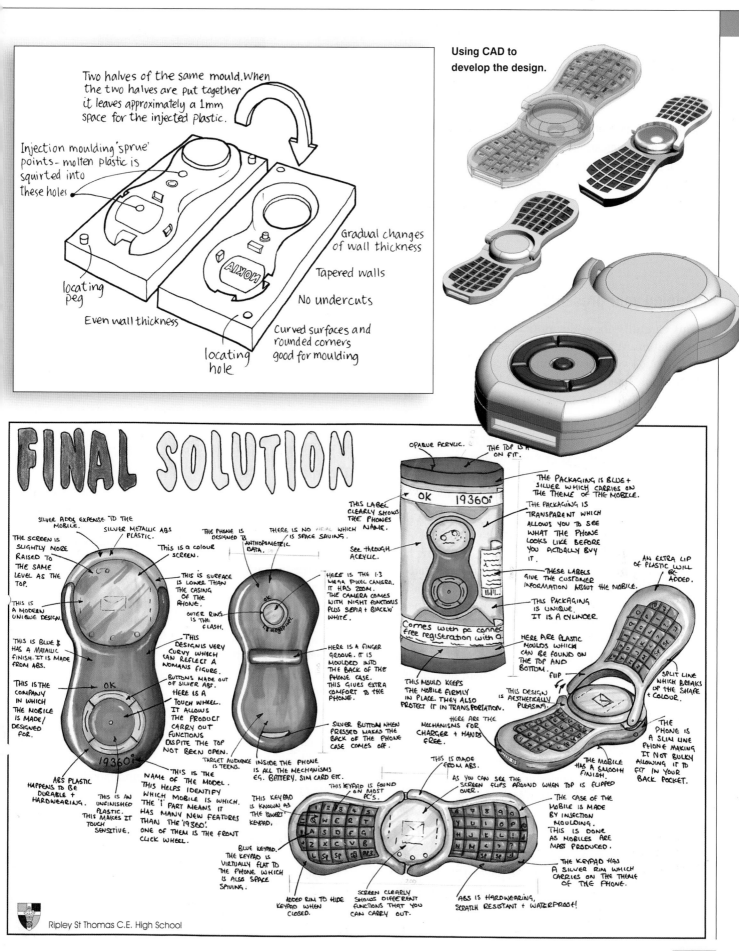

Two halves of the same mould. When the two halves are put together it leaves approximately a 1mm space for the injected plastic.

Injection moulding 'sprue' points - molten plastic is squirted into these holes.

locating peg

Even wall thickness

locating hole

Gradual changes of wall thickness

Tapered walls

No undercuts

Curved surfaces and rounded corners good for moulding

Using CAD to develop the design.

FINAL SOLUTION

SILVER ADDS EXPENSE TO THE MOBILE.

SILVER METALLIC ABS PLASTIC.

THE SCREEN IS SLIGHTLY MORE RAISED TO THE SAME LEVEL AS THE TOP.

THIS IS A COLOUR SCREEN.

THE PHONE IS DESIGNED TO

THERE IS NO AERIAL WHICH IS SPACE SAVING.

THIS IS A MODERN UNIQUE DESIGN.

THIS IS SURFACE IS LOWER THAN THE CASING OF THE PHONE.

ANTHROPOMETRIC DATA.

See through ACRYLIC.

THIS IS BLUE & HAS A METALLIC FINISH. IT IS MADE FROM ABS.

OUTER RING IS THE FLASH.

THIS DESIGN IS VERY CURVY WHICH CAN REFLECT A WOMANS FIGURE.

HERE IS THE 1.3 MEGA PIXEL CAMERA. IT HAS ZOOM. THE CAMERA COMES WITH NIGHT FUNCTIONS PLUS SEPIA + BLACK'N' WHITE.

THIS IS THE COMPANY IN WHICH THE MOBILE IS MADE/ DESIGNED FOR.

OK

BUTTONS MADE OUT OF SILVER ABS.

HERE IS A TOUCH WHEEL. IT ALLOWS THE PRODUCT CARRY OUT FUNCTIONS DESPITE THE TOP NOT BEEN OPEN.

HERE IS A FINGER GROOVE. IT IS MOULDED INTO THE BACK OF THE PHONE CASE. THIS GIVES EXTRA COMFORT TO THE PHONE.

ABS PLASTIC HAPPENS TO BE DURABLE + HARDWEARING.

19360i

THIS IS AN UNFINISHED PLASTIC. THIS MAKES IT TOUCH SENSITIVE.

THIS IS THE NAME OF THE MODEL. THIS HELPS IDENTIFY WHICH MOBILE IS WHICH. THE 'i' PART MEANS IT HAS MANY NEW FEATURES THAN THE '19360'. ONE OF THEM IS THE FRONT CLICK WHEEL.

TARGET AUDIENCE IS TEENS.

INSIDE THE PHONE IS ALL THE MECHANISMS EG. BATTERY. SIM CARD ETC.

SILVER BUTTON WHEN PRESSED MAKES THE BACK OF THE PHONE CASE COMES OFF.

BLUE KEYPAD IS VIRTUALLY FLAT TO THE PHONE WHICH IS ALSO SPACE SAVING.

THIS KEYPAD IS KNOWN AS THE QWERTY KEYPAD.

THIS KEYPAD IS FOUND ON MOST PC'S.

ADDED RIM TO HIDE KEYPAD WHEN CLOSED.

SCREEN CLEARLY SHOWS DIFFERENT FUNCTIONS THAT YOU CAN CARRY OUT.

THIS IS MADE FROM ABS.

ABS IS HARDWEARING, SCRATCH RESISTANT + WATERPROOF!

OPAQUE ACRYLIC.

THE TOP IS A ON FIT.

OK 19360i

THIS LABEL CLEARLY SHOWS THE PHONES NAME.

Comes with pc connect free registration with O

THIS MOULD KEEPS THE MOBILE FIRMLY IN PLACE. THEY ALSO PROTECT IT IN TRANSPORTATION.

THE PACKAGING IS BLUE + SILVER WHICH CARRIES ON THE THEME OF THE MOBILE.

THE PACKAGING IS TRANSPARENT WHICH ALLOWS YOU TO SEE WHAT THE PHONE LOOKS LIKE BEFORE YOU ACTUALLY BUY IT.

THESE LABELS GIVE THE CUSTOMER INFORMATION ABOUT THE MOBILE.

THIS PACKAGING IS UNIQUE. IT IS A CYLINDER.

HERE ARE PLASTIC MOULDS WHICH CAN BE FOUND ON THE TOP AND BOTTOM FLIP.

THIS DESIGN IS AESTHETICALLY PLEASING.

HERE ARE THE MECHANISMS FOR CHARGER + HANDS FREE.

AS YOU CAN SEE THE SCREEN FLIPS AROUND WHEN TOP IS FLIPPED OVER.

AN EXTRA LIP OF PLASTIC WILL BE ADDED.

SPLIT LINE WHICH BREAKS OF THE SHAPE + COLOUR.

THE PHONE IS A SLIM LINE PHONE MAKING IT NOT BULKY ALLOWING IT TO FIT IN YOUR BACK POCKET.

THE MOBILE HAS A SMOOTH FINISH.

THE CASE OF THE MOBILE IS MADE BY INJECTION MOULDING. THIS IS DONE AS MOBILES ARE MASS PRODUCED.

THE KEYPAD HAS A SILVER RIM WHICH CARRIES ON THE THEME OF THE PHONE.

Ripley St Thomas C.E. High School

147

Convince me…

The final stage of your project is to present your ideas to 'In Touch Telecommunications'. What do you need to explain to them, and how are you going to do it?

SENDER   RECEIVER

- **What information do I need to communicate?**
- **What do I want to achieve?**
- **When and where will I be communicating the information, and to whom?**
- **How much detail is needed?**

- **How much do I already know about this?**
- **Why should I be interested?**
- **What's my reaction and response?**
- **What questions am I going to ask?**

Planning your presentation

Think carefully about the **audience** and **purpose** of your presentation. What information about your product and its market do you need to communicate to the directors of 'In Touch Telecommunications'? What you say and show to them must be clear and concise.

Think carefully about the messages you want to send to 'In Touch Telecommunications', and how they will receive and respond to your presentation.

Remember that 'In Touch Telecommunications' asked you to:

- investigate how people use present-day multi-functional mobile phones and PDAs; including how they are **controlled**, **held**, **carried** and **protected**
- suggest the functions mobile communication devices of the future might have, and how they might be used in **specific situations**
- develop **imaginative** design ideas for the general appearance of such a device. Pay particular attention to how the device would be controlled, held, carried and protected.
- suggest what **materials** such a product might be made from and how it might be **mass manufactured**
- produce an appropriate **presentation** of your proposals, including an appearance model of the device and its accessories and a series of explanatory display panels or electronic screens.

And also to:

- create a printed **instruction leaflet** and/or **sales brochure** for your product (see pages 150–151)
- design and make the **packaging** and **point-of-sale** stand for your product (see pages 102–105).

The what
You need to consider:

- what the **main features** of your design are, and which are the most **interesting** and **innovative**
- what things are most likely to convince the directors that your design ideas will be **successful in the market**
- what things the directors are likely to be concerned about and hesitate about investing in.

The how
When you've worked out what you need to communicate, decide how best to explain and show it:

- How can the ideas behind your product be most effectively shown using **models** and **illustrations**?
- How will you communicate your information using **written** or **spoken** words?
- How will aspects of your design be **practically demonstrated**?

IN YOUR PROJECT

- Start by deciding on the main subject of each panel or screen and state this clearly with a main title.
- Work out the key points you want to get across on each panel or screen and make sure these come over clearly.

Presentation Matters

Getting the point across

Before the development of electronic means of communication, design proposals were always presented to clients using a series of display boards. In many situations this still happens. However, these are increasingly supplemented by the use of PC screen-based presentation programs.

Keep it simple...

Whichever methods you use, it's important to consider the design of the display panels or screens. They need to be clearly set out and visually stimulating. You need to try to develop a style of your own that reflects the individuality and creativity of your design ideas: don't just do the same as everyone else, and avoid using familiar electronic 'wizard' templates. But the most important piece of advice is to keep it all as simple as possible.

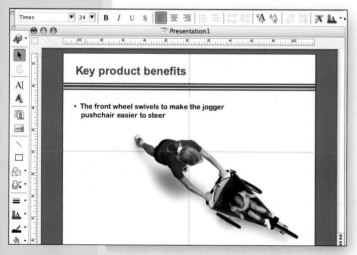

Design checklist

- Limit the number of bullet points on each screen.
- Be very sparing with animated transition effects.
- Avoid the use of cartoon clipart – use a good photograph instead.
- Colour evokes an emotional response – use it carefully (see page 100).
- Try not to use more than two different fonts – ideally sans serif ones (see page 97). Make sure their size and weight provide just the right amount of legibility when viewed from a distance (or close-up).
- Use the opportunity to add video and audio – but not too much.

Powerful presentations

Here are some ways to make your presentation more interesting and memorable:

- Use **graphs** to show the changing relationship between two factors, one of which is often time.

- Use **bar charts** and **histograms** to compare data. Divide the bars to show more complex information. Add colour, both solid and graduated, textured, and/or images relevant to the subject matter.

- A **pie chart** shows information as a proportion of the whole. They can be given much greater visual impact by adding colour and displaying in 3D. Showing a segment partly removed can also be very effective.

- **Flow charts** communicate sequences of events in a process. The distance shown between the stages can be made to represent physical space, or time.

- **Network diagrams** show the relationships between objects, people or ideas.

Designers use 3D presentation models to help communicate their ideas to their clients. Sometimes the models work, but their main purpose is to show what a product or space will look and feel like in real life. Photographing a model of a product being used by someone in the correct setting helps to make it even more convincing.

■ ACTIVITY

Experiment by trying out some of the following suggestions:

- Break up a plain white surface by using strips of colour and texture.
- Position titles and explanatory diagrams on unusually-shaped cut-outs.
- Develop a co-ordinated colour scheme that works across all your panels or screens.
- Montage/overlap photographs together. Show close-ups of the product.
- Cut out images completely from their backgrounds.
- Devise a numbering system, logo and perhaps a border.
- Use extra or fewer screens.

Information design

Designing successful instructional leaflets and promotional brochures involves achieving the best layout of text, illustrations and graphic devices on the chosen format, all within the available budget.

'It's easy to use...'

Unfortunately, products can't speak for themselves, at least not yet. They can't tell you how they should be used, so an **instruction manual** is needed. These include things like what you need to know about assembly, daily use, necessary maintenance, safety warnings and what to do if it doesn't work.

Manuals use a wide variety of words and visual clues to get users to pull or push and twist or turn, rotate and insert. They show which parts are which, what the different displays and controls do, how they all fit together, how far apart things should be, what order to do things in, as well as what to check and adjust.

The problem is that all this becomes so complicated that most people can not be bothered to consult the manual – they just want to start using the product. As a result, they don't put it together properly, or never get to discover all the clever things it can do, if only they knew how.

'It's simply the best...'

Promotional leaflets and **brochures** are another common form of information design. Instead of communicating how to use a product, their purpose is to help persuade you to buy one in the first place.

Advertisements are there to inform you that a product exists and suggest you might want one. The job of the brochure is to provide more information to help inform your decision.

Brochures tell you more about its features – what it will do for you and how it will improve your life – and some more technical information about its performance. Sometimes they will compare the product favourably with other makes, and include 'testimonials' from satisfied users or endorsements from other companies and organisations.

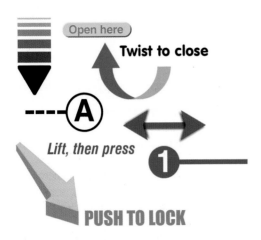

Are these sorts of instructions and graphics helpful or confusing?

■ Layout design checklist

Text

- What information is best provided in written form?

- Which sections of text will be used as:
  - titles and headings
  - main (or 'body') text and any 'small print'
  - captions?

- What size and style of typefaces will be most appropriate for each section?
  - Serif or san serif.
  - Traditional or decorative.
  - Small or large.
  - Thick or thin.
  - Black, white, or coloured.

(See also page 97.)

ABC 123
abc 123 abc 123

Illustrations

- What information will be best provided by means of graphs and charts, plans, illustrative drawings, photographs, etc?

- Will artwork be realistic, diagrammatic or impressionistic?

- What size will illustrations be?

- Will they be in colour, or black and white?

- What graphic devices, i.e. lines or flat areas of colour might be effective?

- What text will illustrations need to be near to?

Format

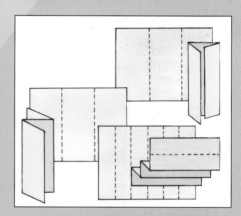

- What size and type of paper will be used, and how will it be folded and/or fastened together?

- How will these decisions affect the printing and production costs?

Layout

- How will the text and the illustrations be best arranged in the format?

Grid references

Behind every printed surface there is a grid structure. The grid defines the height, width and position of the margins, columns and headings. Everything lines up with something. This makes the information easier to read and understand, and more pleasing to the eye.

The first stage of any layout design is the development of the grid.

Thumbnails

To create a grid, you will need to explore some ideas in rough using what are known as **thumbnail** sketches.

These are small, quick drawings that each represent the possible layout of the page.

As you develop the layout, experiment with the possibilities for:

- having one, two, three or more columns
- the size of the margins between the columns
- how the grid will be divided horizontally
- what graphic devices, such as lines, tints and slabs of colour, could be used
- where a running head and/or any page numbers will be placed
- the size and starting position of main headings.

On the desktop

When the basic approach to the layout has been decided, the grid can be set up and refined using a DTP program. Try to avoid using the ready-made template 'wizards' – as a designer you should be creating something original.

When the underlying grid has been completed, start to place the main body text, headings, captions, photographs and illustrations. Before the use of DTP these elements were literally cut out and pasted down. Now they can be easily and quickly moved around on screen until the best layout is achieved.

■ ACTIVITY

Obtain a product brochure and place a sheet of tracing paper over a page of text.

Using a pencil and ruler, draw in the vertical and horizontal lines of the main columns of body text to reveal the underlying layout grid.

Measure the widths of the columns and margins and distance between the horizontal lines.

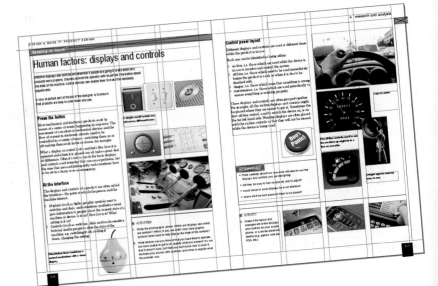

- **My product is...**
- **The design is aimed at...**
- **Its key features are...**
- **What makes it unique is...**
- **It will be easy and cheap to mass-produce because...**

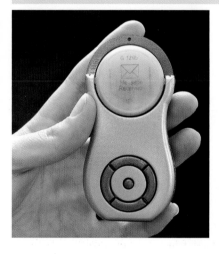

Final testing

Assemble all the things you have created to present to 'In Touch Telecommunications' – models, displays, instructions, brochures, packaging and/or point-of-sale units.

Pitch your design idea to one of the other groups in your class (i.e. who are unfamiliar with your designs) as if they were representing 'In Touch Telecommunications'. Initially, you only have 30 seconds to explain your basic design proposal, so make sure you tell them very clearly and concisely what it is and why it's a product they should invest in.

Then, give the group no more than five minutes to look at your models and other presentation materials and ask you questions about it, and express any concerns or criticisms they might have.

Finally, ask them some specific questions to discover how well they have understood exactly:

- what the product being offered is
- who the target market is, and why they might benefit from and want such a product
- its particular innovative features
- its potential for large-scale production.

Final evaluation

Review your work on the project as a whole. In particular, consider the following:

- If you were re-designing the product or any of the presentation items, what might you do differently?
- If you were going to develop your existing designs further, what problems would you need to solve?

Also comment on the effectiveness of your final pitch to 'In Touch Telecommunications':

- How well did they understand your design proposals?
- Did you agree with any criticisms or suggestions they made?

Finally, comment on how well you worked:

- Which parts of the process did you find the easiest, most challenging, etc?
- To what extent did you manage to build on your strengths and improve your weaknesses?
- What have you learned most during the project?

Examination questions

Q2

You should spend about two hours answering the following questions. You will need some plain A4 paper, basic drawing equipment and colouring materials. You are reminded of the need for good English and clear presentation in your answers.

1. This question is about **designing new products.** *See pages 16–23, 132–133. Spend about 20 minutes on this question.*

a)   Many designers are now household names. Name *two* designers from the past or present and describe what they are famous for. *(6 marks)*

b)   The Sony Walkman is an example of 'market pull'. Name a product that you feel has been developed because of market pull and explain why you believe this to be the case. *(4 marks)*

c)   Textured vegetable protein (TVP) has been created from experiments with Soya beans. This is an example of 'technology push'. Name a product that you feel has been developed because of technology push and explain why you believe this to be the case. *(4 marks)*

d)   New products are often developed from previous products with similar functions. This is known as product evolution. An iron is a good example of this. Using notes and sketches, show *one* example of a different product that has developed over time. *(4 marks)*

2. This question is about **developing design ideas.** *See pages 56–57. Spend about 15 minutes on this question.*

New ideas come from many different sources but nature and mathematics are two rich areas.

a)   Using nature as a starting point, draw an idea for a new piece of jewellery. Add notes to explain what materials and processes you would use and add colour and tone to make your drawing look more realistic. *(8 marks)*

b)   Using geometry as a starting point, draw an idea for a new kettle. Add notes to explain what materials and processes you would use and add colour and tone to make your drawing look more realistic. *(8 marks)*

3. This question is about CAD-CAM. *See pages 10–11, 142–143. Spend about 30 minutes on this question.*

a)   Computer-aided design (CAD) is used both in schools and in industry. Explain *three* methods you could use to get an illustration on to a design for a child's book. *(6 marks)*

b)   List *three* computer-aided manufacturing (CAM) processes and describe a good use for each process. *(6 marks)*

c)   CAM is used a great deal in manufacturing industry. Describe *three* examples of its application. *(6 marks)*

d)   Give *three* advantages of using CAM over traditional manufacturing methods. *(3 marks)*

e)   Give *three* disadvantages of using CAM over traditional manufacturing methods. *(3 marks)*

f)   Using notes and sketches, explain how you could use CAM to make a high-quality prototype to show to a client. You could base this on your recent project. *(6 marks)*

4. This question is about manufacturing in quantity. *See pages 120–125. Spend about 35 minutes on this question.*

You have been asked by the parents' association to design and manufacture a small novelty toy or game to sell at the school summer fair. 300 identical products are required. Your design will need to be simple to manufacture in quantity and can be in any material of your choice.

a) Produce at least *three* ideas for a suitable gift which can be manufactured in quantity using your Design and Technology facilities. Add notes to explain the materials and processes which could be used.
(9 marks)

b) Select the best idea and, using notes and sketches, explain in detail how this will be manufactured using Design and Technology facilities.
(6 marks)

c) Explain how you will ensure that every gift is identical and of a high quality. *(3 marks)*

d) What name is given to the process of ensuring every product is of identical high quality? *(2 marks)*

e) Explain how you will organise a team of *three* students to produce the 300 products. *(6 marks)*

f) Explain how you will ensure that every student in your production team is safe at all times. *(6 marks)*

g) What name is given to the activity of ensuring everyone is safe before an activity starts? *(2 marks)*

5. This question is about assembling products. *See pages 32–37. Spend about 10 minutes on this question.*

a) Products are often assembled from several parts. Explain why this might be the case. *(2 marks)*

b) Using notes and sketches, explain in detail *one* assembly method you have used during your course. *(4 marks)*

c) Many products are designed for self-assembly. Give *two* advantages of producing self-assembly products.
(2 marks)

d) Using notes and sketches, explain in detail *one* non-permanent fastening found on products.
(3 marks)

6. This question is about how computers are used in designing and manufacturing. *See pages 68–69. Spend about 10 minutes on this question.*

a) Name *one* design application for using computers and explain the advantages of this method over traditional methods. *(5 marks)*

b) Name *one* manufacturing application for using computers and explain the advantages of this method over traditional methods.
(5 marks)

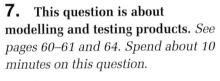

7. This question is about modelling and testing products. *See pages 60–61 and 64. Spend about 10 minutes on this question.*

Product designers need to model and test their ideas.

a) Choose *one* product from the following list and briefly explain how you could model it. *(3 marks)*

Decorative cake	Timing device
Hat	Mobile phone
Board game	Waistcoat
Lamp	Photo frame

b) Once a designer has made a model of a product, tests are often needed. Describe *one* test that could be used to evaluate the success of the design. *(2 marks)*

c) Symbols are often found on packaging to tell you that the product has been tested against a set of standards. Draw *one* of these symbols and name the organisation it represents. *(4 marks)*

Crafty kits

A major supermarket chain plans to produce a range of new and imaginative craft and hobby kits to sell in their leisure section. Each kit will need to be appropriately packaged and contain everything the user will need to complete the task, including detailed instructions.

What ideas for suitable products can you come up with?

The task

The supermarket's market research department has suggested the following range of kits that might be suitable for their customers:

- Textile crafts
- Cookery
- Greeting cards
- Construction
- Electronics
- Scale models
- Candle making
- Jewellery

However, they would also welcome other ideas of your own. Whichever type of kit you choose, you will need to establish a clear design brief aimed at a specific age range, based on your own market research.

You will need to fully understand the craft or hobby you are designing for – this may involve you learning new skills yourself.

You will also need to be able to communicate to someone else the skills required to complete the task, using clear and concise illustrations, words and numbers.

First thoughts

There is already a wide range of craft and hobby kits available in the shops. What unusual and exciting ideas can you come up with that are going to interest and impress the supermarket? Something a bit different is needed. Think just as much about who the user will be and the circumstances or situations in which he or she (or even a group of people) will be making and using the product.

Find out more

Investigate what types of craft and hobby kits already exist on the market and conduct a detailed product analysis of at least two. Make sure you include the packaging and instructions in your study.

If possible, talk to some people who use such kits. Why do they buy them? What makes them particularly satisfying, or frustrating to assemble?

Specify

Decide on a target market for your kit and establish your initial design criteria.

Inspiration

How can you make your kits as visually stimulating as possible to encourage people to buy them and to add to their enjoyment of making and owning them? What shapes, forms, colours and textures will be particularly appropriate?

Modelling and prototypes

Develop a range of ideas for the kit and how it could be packaged. This will include a great deal of modelling, so do keep a record of all the work you do. Take plenty of photographs of your design ideas as they progress.

Planning and manufacturing

A fully detailed design is needed that will provide enough information for someone else to manufacture a similar prototype. This should match your design specification.

You will need to complete the kit yourself in order to photograph the various stages of assembly. This may mean that you will need to manufacture more than one set of components/ingredients.

You will need to show all of your planning, any modifications made and the quality assurance procedures you will need to follow.

Your final design proposal will need to be a complete kit as you would imagine it on sale in the supermarket. Remember, this will need to include the packaging and instructions, as well as the materials and components.

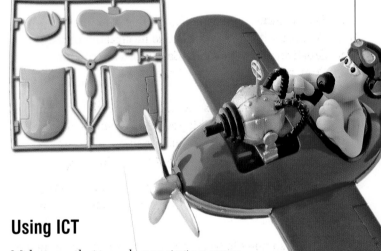

Using ICT

Make sure that you demonstrate a range of graphical =and ICT communication skills appropriate to this task. The specific materials you are working with will determine your potential use of CAD-CAM.

Final evaluation

You will be evaluating different aspects of the kit throughout the project. However, you should aim to conduct some consumer testing with the completed kit.

■ Are the instructions clear?
■ Is everything you need included?
■ Are there any changes needed to allow your design to go into full production?

Project suggestions [2]

Making light work

A lighting company needs design ideas for a new floor or table lamp.
Can you solve the problems involved in designing a new lighting product?

**Look out for familiar
examples of things
that stack in some way.**

The task

You have been asked by a lighting company called 'IDEA'
to develop a new floor or table lamp that uses a 12-volt
power unit.

The product must be capable of being sold in a self-
assembly format and will need to be suitably packaged.
There will need to be simple assembly instructions.

The company's marketing department is particularly keen
to have a concept that they can call the 'Tower', based on
the idea of a stack of identical shapes with spacers in
between. The 12-volt light unit must fit into the base.

SAFETY FIRST!

On no account should you use
AC mains power at any time
while working on this task.

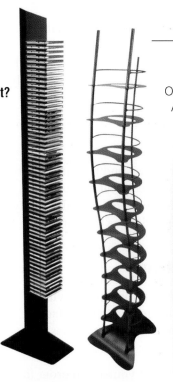

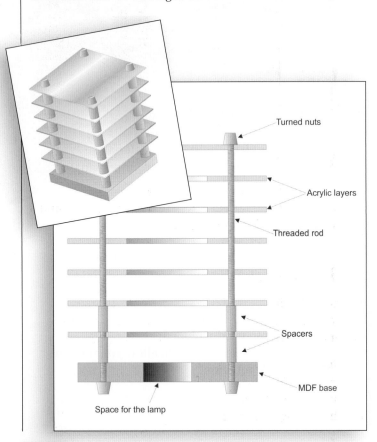

Turned nuts

Acrylic layers

Threaded rod

Spacers

MDF base

Space for the lamp

Find out more

Make a study of free-standing floor and table lighting
devices.

- Identify the sort of places where they are intended
  to be used (e.g. living room, bedroom, etc.), and the
  sort of people who might buy them. What are their
  preferences for materials, colours, patterns and
  textures likely to be?
- Create a mood board to indicate the sort of style and
  environment the lamp you intend to design would be
  used in.

Undertake a detailed product analysis of a specific
lighting device, ideally with its original packaging and
assembly instructions.

Develop the design

As you develop your design idea there are some specific problems you will need to solve.

1 The cable

The cable from the lamp unit will need to be routed through the base of the tower. That might mean altering the base so it can be drilled and the cable brought out of the side, or it might mean fitting feet to the base so there is a space for the cable underneath.

2 The spacer

There will need to be some form of spacer made to separate each layer in the design. This may mean manufacturing a cutting jig to ensure that each one is the same. What other methods could be considered?

3 The material

Consider what material each layer will be made from and how the light will pass through the tower. Is there a need to have holes in each layer?

4 Construction

What method will be used to fasten all of the pieces together? Could some form of threaded rod system be made attractive? If so, what type of nuts would you use? Can these be made especially for your design?

In your development you need to demonstrate that you have approached each of these problems in an imaginative way, i.e. not just choosing the obvious solution. Your design sheets will need to show that you have come up with different:

- possible ways of routing the cable
- shapes for the spacers
- materials that allow light to pass through
- construction methods.

Soft modelling

After you've done some initial sketching on paper, move towards experimenting with simple or scrap materials. You might want to work at a smaller scale to begin with. These will not be working models, but will allow you to consider how the cable, spacers, materials and construction methods will work together in three dimensions.

Make sure you take regular photographs of the models you make.

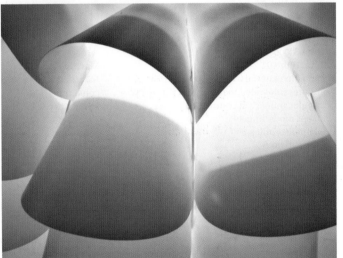

Working prototype

You will need to make a full-size working prototype, ideally using the actual materials and fittings that would be used in the production model.

Final presentation

Prepare suitable materials to send to 'IDEA' for their consideration. These will need to be completely self-explanatory. Will they be able to easily understand:

- what it looks like
- how it fits together
- what it's made from
- who the target market is?

Index